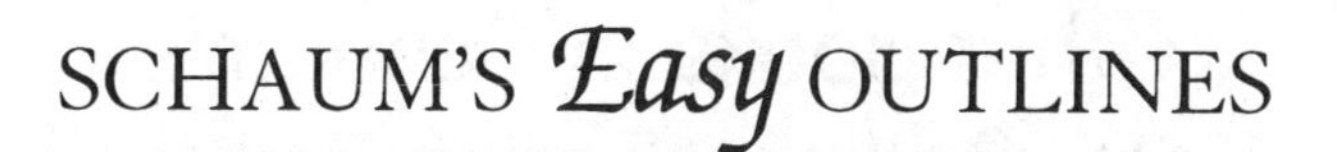
SCHAUM'S Easy OUTLINES

GEOMETRY

Other Books in Schaum's Easy Outlines Series Include:

Schaum's Easy Outline: Calculus
Schaum's Easy Outline: College Algebra
Schaum's Easy Outline: College Mathematics
Schaum's Easy Outline: Mathematical Handbook of Formulas and Tables
Schaum's Easy Outline: Statistics
Schaum's Easy Outline: Biology
Schaum's Easy Outline: College Chemistry
Schaum's Easy Outline: Human Anatomy and Physiology
Schaum's Easy Outline: Organic Chemistry
Schaum's Easy Outline: Physics
Schaum's Easy Outline: Principles of Accounting
Schaum's Easy Outline: Programming with C++
Schaum's Easy Outline: Programming with Java
Schaum's Easy Outline: French
Schaum's Easy Outline: German
Schaum's Easy Outline: Spanish
Schaum's Easy Outline: Writing and Grammar

SCHAUM'S *Easy* OUTLINES

GEOMETRY

BASED ON SCHAUM'S
Outline of Geometry
BY BARNETT RICH
REVISED BY PHILIP A. SCHMIDT

ABRIDGEMENT EDITOR:
GEORGE J. HADEMENOS

SCHAUM'S OUTLINE SERIES

McGRAW-HILL

New York San Francisco Washington, D.C. Auckland Bogotá Caracas Lisbon London Madrid Mexico City Milan Montreal New Delhi San Juan Singapore Sydney Tokyo Toronto

BARNETT RICH earned a Ph.D. from Columbia University and a J.D. from New York University. He was a founder of the New York City High School of Music and Art and was chairman of Mathematics at the Brooklyn Technical High School. He also taught at City University of New York and Columbia.

PHILIP A. SCHMIDT has a B.S. from Brooklyn College and both M.A. and Ph.D. degrees from Syracuse University. He is Associate Provost for Academic Services at Berea College in Kentucky and previously Dean of the School of Education at SUNY at New Paltz.

GEORGE J. HADEMENOS has taught at the University of Dallas and done research at the University of Massachusetts Medical Center and the University of California at Los Angeles. He holds a B.S. from Angelo State University and both M.S. and Ph.D. degrees from the University of Texas at Dallas. He is the author of several books in the Schaum's Outline series.

1 2 3 4 5 6 7 8 9 0 DOC DOC 0 9 8 7 6 5 4 3 2 1

ISBN 0-07-136973-2

McGraw-Hill

*A Division of The **McGraw-Hill** Companies*

Contents

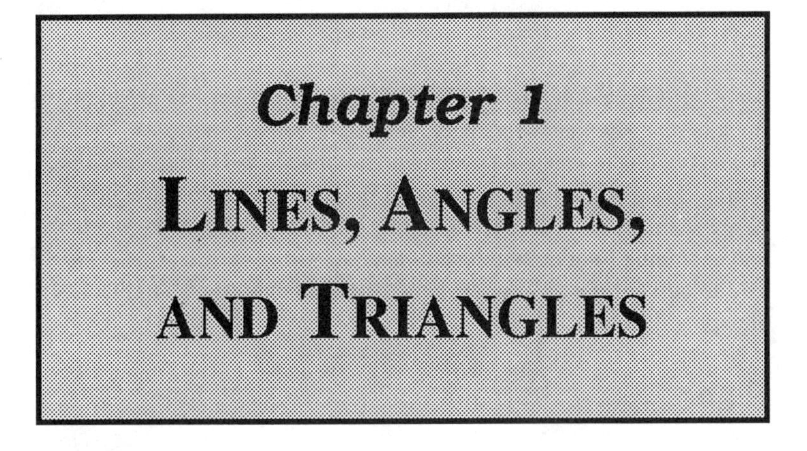

Chapter 1
Lines, Angles, and Triangles

In This Chapter:

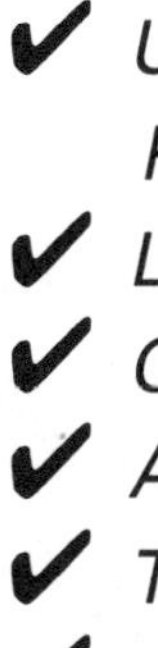

- ✔ *Undefined Terms of Geometry: Point, Line, and Plane*
- ✔ *Line Segments*
- ✔ *Circles*
- ✔ *Angles*
- ✔ *Triangles*
- ✔ *Pairs of Angles*

Undefined Terms of Geometry: Point, Line, and Plane

Point

A *point*, represented by a dot, has position only. It has no length, width, or thickness.

Line

A *line*, designated by the symbol, $\overleftrightarrow{AB}$, has length but has no width or thickness. A line may be straight, curved, or a combination of these. A *straight line* is generated by a point moving always in the same direction. A straight line may be extended in either direction indefinitely. A *ray*, written as $\overrightarrow{AB}$, is the part of a straight line beginning at a given point and extending limitlessly in one direction.

A *curved line* is generated by a point moving in a continuously changing direction.

Plane

A *plane* has length and width but no thickness. A plane is a surface such that a straight line connecting any two of its points lies entirely in it.

Line Segments

A straight line segment, designated by $\overline{AB}$, is the part of a straight line between two of its points, including the two points.

If a line segment is divided into parts:

1. The length of the whole line segment equals the sum of the lengths of its parts. Note that the length of $\overline{AB}$ is designated AB.
2. The length of the whole line segment is greater than the length of any part.
3. Two line segments having the same length are said to be *congruent*. Thus, if $AB = CD$, then $\overline{AB}$ is congruent to $\overline{CD}$, written $\overline{AB} \cong \overline{CD}$.

If a line segment is divided into two equal parts:

1. The point of division is the midpoint of the line segment.
2. A line that crosses at the midpoint is said to bisect the segment.

Because $AM = MB$ in Fig. 1-1, M is the midpoint of $\overline{AB}$, and $\overline{CD}$ bisects $\overline{AB}$. Equal line segments may be shown by crossing them with the same number of strokes. Note that $\overline{AM}$ and $\overline{MB}$ are crossed with a single stroke.

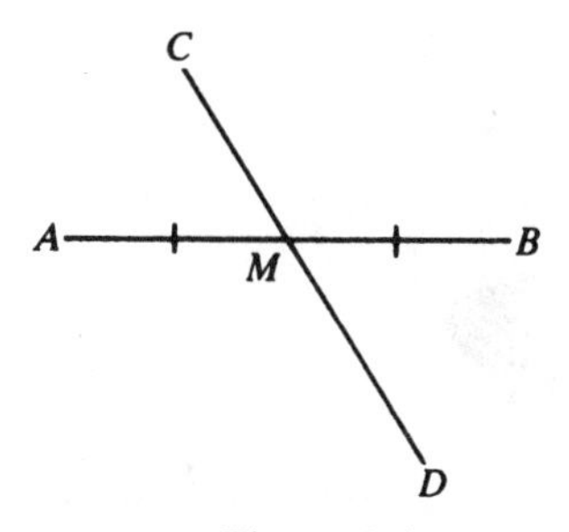

Figure 1-1

3. If three points A, B, and C lie on a line, then they are *collinear*. If A, B, and C are collinear and $AB + BC = AC$, then B is between A and C. [See Fig. 1-2.]

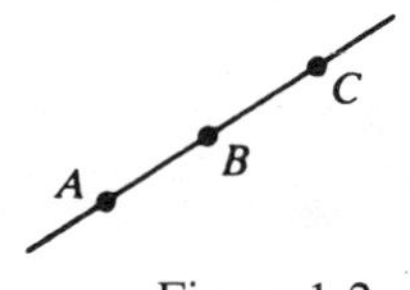

Figure 1-2

Circles

A *circle* is the set of all points in a plane that are the same distance from the *center*.

The *circumference* of a circle is the distance around the circle. It contains degrees (360°).

A *radius* is a segment joining the center of a circle to a point on the circle (see Fig. 1-3).

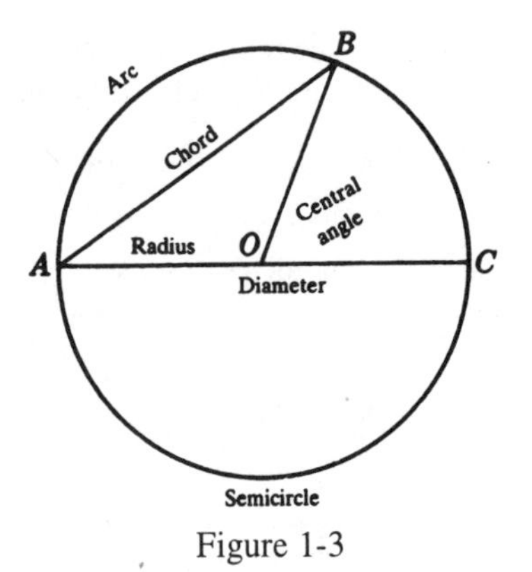

Figure 1-3

From the definition of a circle, it follows that the radii of a circle are congruent.

A *chord* is a segment joining any two points on a circle.

A *diameter* is a chord through the center of the circle; it is the longest chord and is twice the length of a radius.

An *arc* is a continuous part of a circle. An arc of measure 1° is 1/360th of a circumference.

A *semicircle* is an arc measuring one-half the circumference of a circle and thus contains 180°. A diameter divides a circle into two semicircles.

A *central angle* is an angle formed by two radii.

Congruent circles are circles having congruent radii.

Angles

An *angle* is the figure formed by two rays with a common end point. The rays are the *sides* of the angle, while the end point is its *vertex*. The symbol for angle is $\angle$ or $\measuredangle$. Thus, $\overrightarrow{AB}$ and $\overrightarrow{AC}$ are the sides of the angle shown in Fig. 1-4, and A is its vertex.

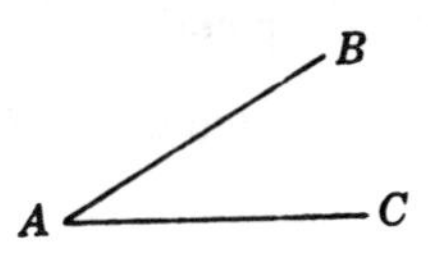

Figure 1-4

Measuring the Size of an Angle

The size of an angle depends on the extent to which one side of the angle must be rotated, or turned about the vertex, until it meets the other side. We choose degrees to be the unit of measure for angles. The measure of an angle is the number of degrees it contains. We will write $m\angle A = 60°$ to denote that angle A measures 60°.

The protractor in Fig. 1-5 shows that $\angle A$ measures 60°. If $\overrightarrow{AC}$ were rotated about the vertex A until it met $\overrightarrow{AB}$, the amount of turn would be 60°.

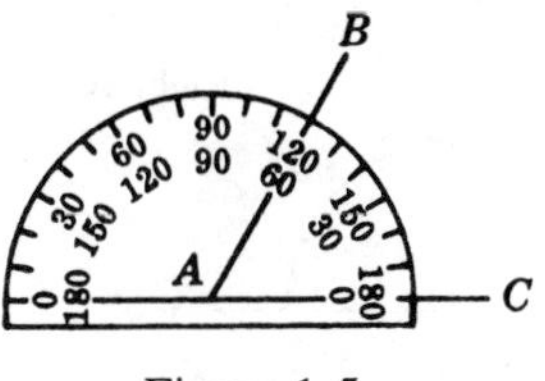

Figure 1-5

In using a protractor, be sure that the vertex of the angle is at the center and that one side is along the 0° – 180° diameter.

The size of an angle *does not* depend on the lengths of the sides of the angle. The size of $\angle B$ in Fig. 1-6 would not be changed if its sides $\overrightarrow{AB}$ and $\overrightarrow{BC}$ were made shorter or longer.

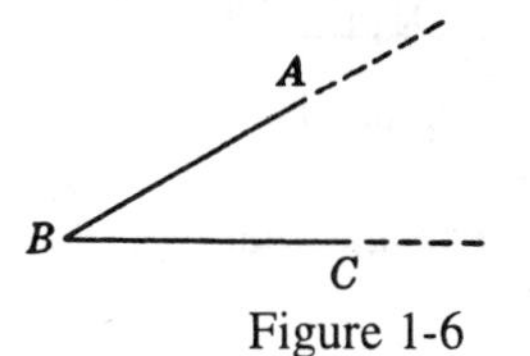

Figure 1-6

To measure angles with more precision, we divide 1° into 60 equal parts, called *minutes*. Thus, 1° = 60 minutes (60′) and 1′ = 60 seconds (60″).

Kinds of Angles

1. *Acute angle*: An acute angle is an angle whose measure is less than 90° (see Fig. 1-7).

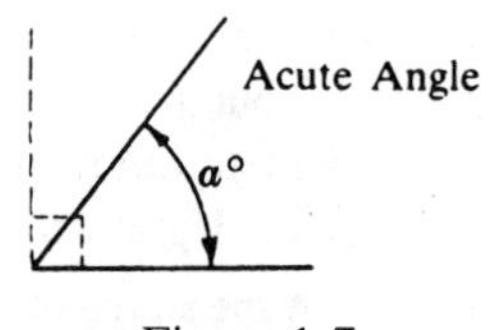

Figure 1-7

2. *Right angle*: A right angle is an angle that measures 90° (see Fig. 1-8).

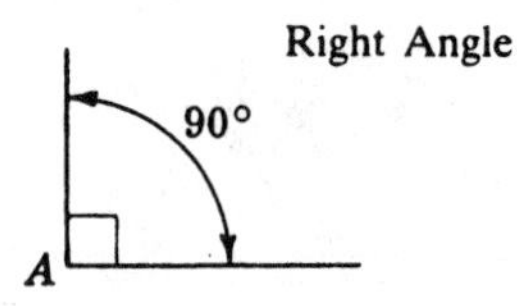

Figure 1-8

3. *Obtuse angle*: An obtuse angle is an angle whose measure is more than 90° and less than 180° (see Fig. 1-9).

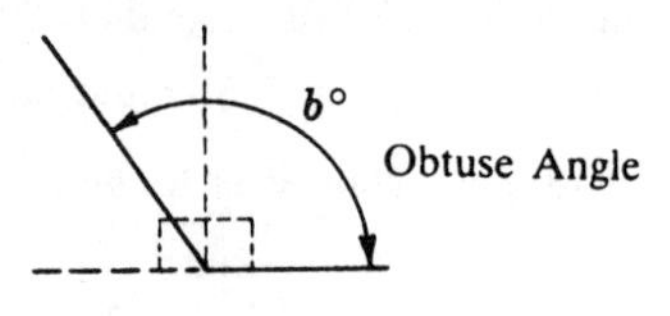

Figure 1-9

4. *Straight angle*: A straight angle is an angle that measures 180° (see Fig. 1-10).

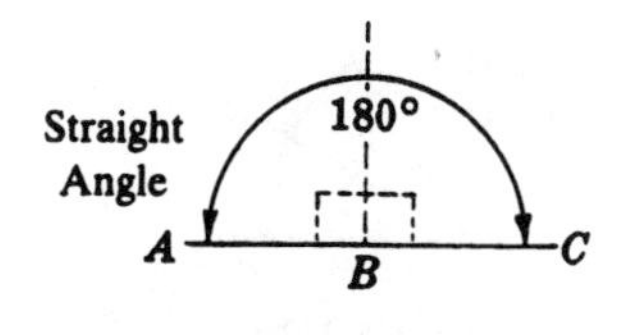

Figure 1-10

Note that the sides of a straight angle lie in the same straight line. But do not confuse a straight angle with a straight line.

5. *Reflex angle*: A reflex angle is an angle whose measure is more than 180° and less than 360° (see Fig. 1-11).

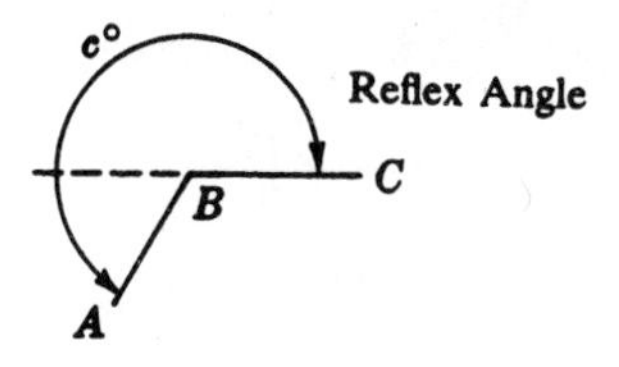

Figure 1-11

Additional Angle Facts

1. *Congruent angles* are angles that have the same number of degrees. In other words, $m\angle A = m\angle B$, then $\angle A \cong \angle B$. Thus, in Fig. 1-12, right (rt.) $\angle A \cong$ rt. $\angle B$ since each measures 90°.

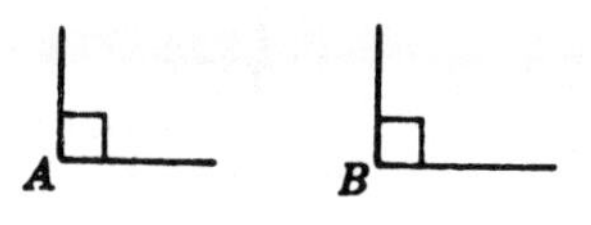

Figure 1-12

2. A line that bisects an angle divides it into two congruent parts. Thus, in Fig. 1-13, if $\overline{AD}$ bisects $\angle A$, then $\angle 1 \cong \angle 2$.

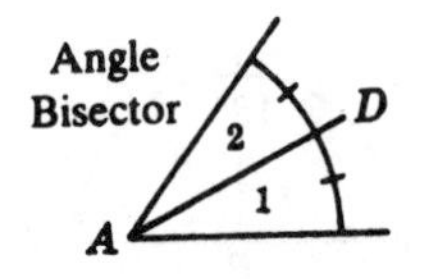

Figure 1-13

3. *Perpendiculars* are lines, rays, or segments that meet at right angles. The symbol for perpendicular is $\perp$. Thus, in Fig. 1-14, $\overline{CD} \perp \overline{AB}$, so right angles 1 and 2 are formed.

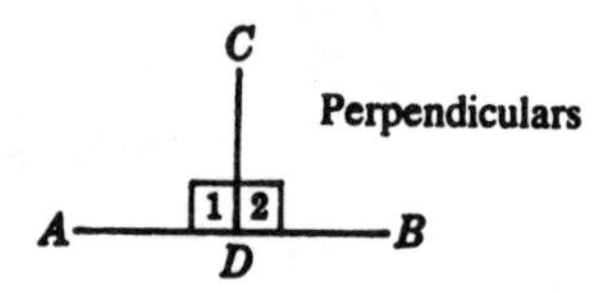

Figure 1-14

4. A *perpendicular bisector* of a given segment is perpendicular to the segment and bisects it. Thus, in Fig. 1-15, $\overline{GH}$ is the $\perp$ bisector of $\overline{EF}$; thus $\angle 1$ and $\angle 2$ are right angles and M is the midpoint of $\overline{EF}$

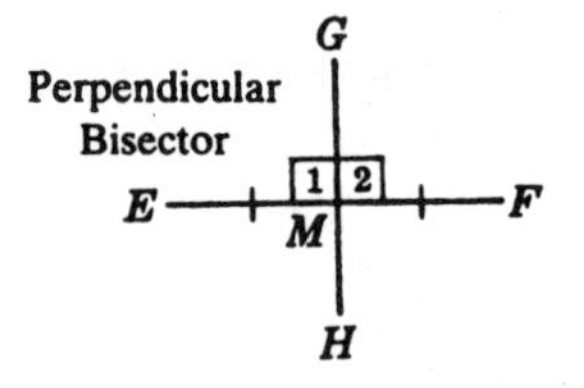

Figure 1-15

Triangles

A *polygon* is a closed plane figure bounded by straight line segments as sides. Thus, Fig. 1-16 is a polygon of five sides, called a *pentagon*; it is named pentagon *ABCDE*, using its letters in order.

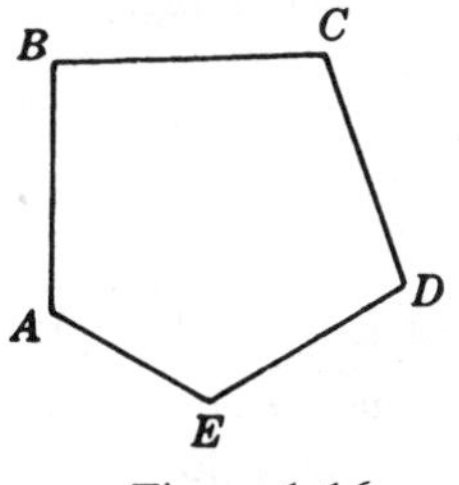

Figure 1-16

A *triangle* is a polygon having three sides. A *vertex* of a triangle is a point at which two of the sides meet. The symbol for triangle is Δ. Thus, the triangle in Fig. 1-17 may be named $\Delta\ ABC$ or ΔI; its sides are $\overline{AB}$, $\overline{AC}$, and $\overline{BC}$; its vertices are A, B, and C; its angles are $\angle A$, $\angle B$, and $\angle C$.

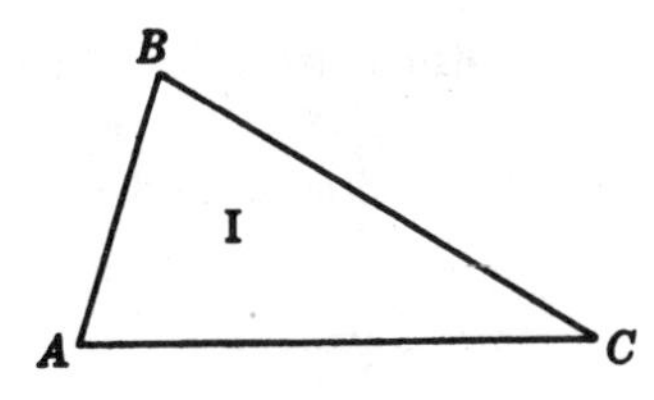

Figure 1-17

Classifying Triangles

Triangles are classified according to the equality of the lengths of their sides or according to the kind of angles they have.

Triangles According to the Equality of the Lengths of Their Sides

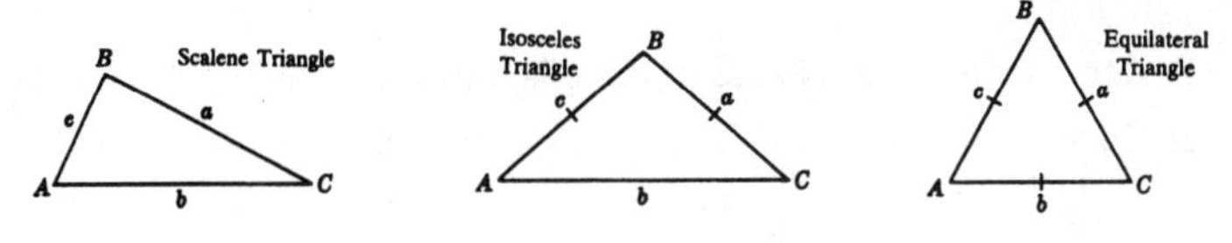

Figure 1-18

1. *Scalene triangle*: A scalene triangle is a triangle having no congruent sides. In scalene triangle ABC, $a \neq b \neq c$. The small letter used for the length of each side agrees with the capital letter of the angle opposite it.

2. *Isosceles triangle*: An isosceles triangle is a triangle having at least two congruent sides. In isosceles triangle ABC, $a = c$. These equal sides are called the *legs* of the isosceles triangle; the remaining side is the base b. The angles on either side of the base are the *base angles*; the angle opposite the base is the *vertex angle*.

3. *Equilateral triangle*: An equilateral triangle is a triangle having three congruent sides. In equilateral triangle ABC, $a = b = c$.

An equilateral triangle is also an isosceles triangle.

Triangles According to the Kind of Angles

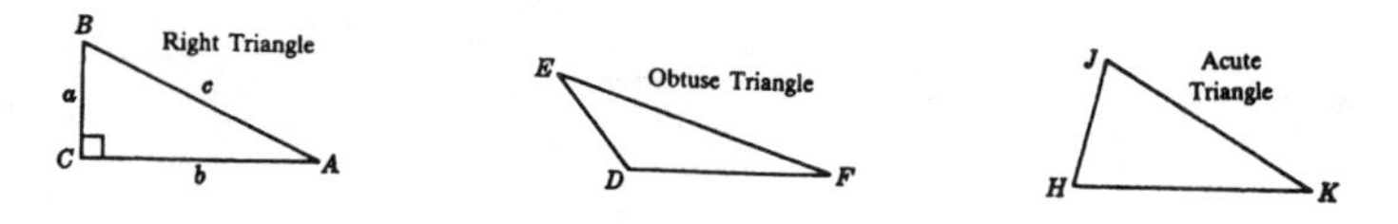

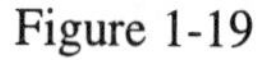
Figure 1-19

1. *Right triangle*: A right triangle is a triangle having a right angle. In right triangle ABC, $\angle C$ is the right angle. Side c opposite the right angle is the *hypotenuse*. The perpendicular sides, a and b, are the *legs* or *arms* of the right triangle.

2. *Obtuse triangle*: An obtuse triangle is a triangle having an obtuse angle. In obtuse triangle DEF, $\angle D$ is the obtuse angle.

3. *Acute triangle*: An acute triangle is a triangle having three acute angles. In acute triangle HJK, $\angle H$, $\angle J$, $\angle K$ are acute angles.

Special Lines in a Triangle

1. *Angle bisector of a triangle*: An angle bisector of a triangle is a segment or ray that bisects an angle and extends to the opposite side. Thus, $\overline{BD}$, the angle bisector of $\angle B$ in Fig. 1-20, bisects $\angle B$, making $\angle 1 = \angle 2$.

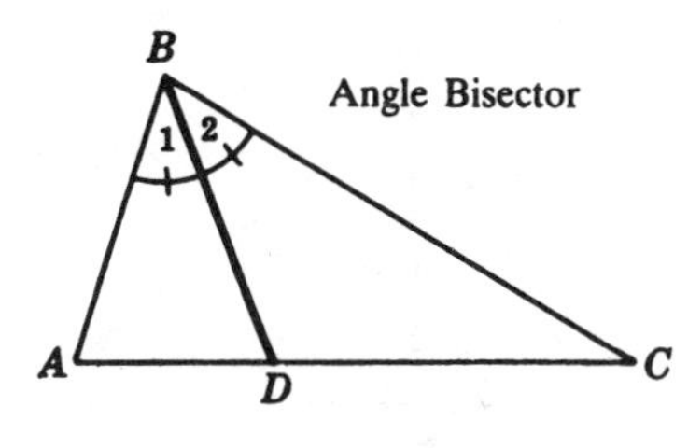

Figure 1-20

2. *Median of a triangle*: A median of a triangle is a segment from a vertex to the midpoint of the opposite side. Thus, $\overline{BM}$, the median to $\overline{AC}$ in Fig. 1-21, bisects $\overline{AC}$, making $AM = MC$.

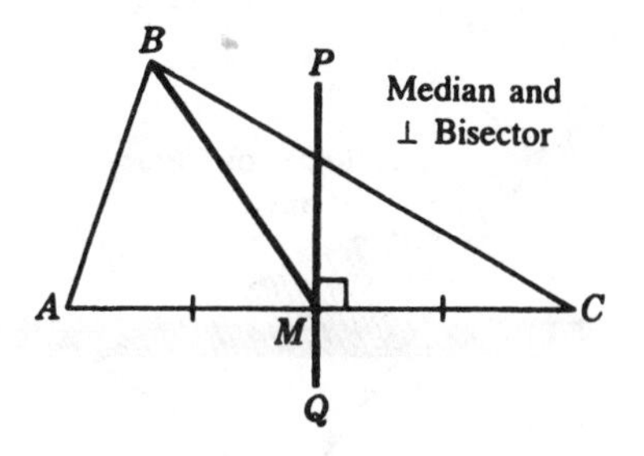

Figure 1-21

3. *Perpendicular bisector of a side*: A perpendicular bisector of a side of a triangle is a line that bisects and is perpendicular to a side. Thus, $\overleftrightarrow{PQ}$, the perpendicular bisector of $\overline{AC}$ in Fig. 1-21, bisects $\overline{AC}$ and is perpendicular to it.

4. *Altitude to a side of a triangle*: An altitude of a triangle is a segment from a vertex perpendicular to the opposite side. Thus, $\overline{BD}$, the altitude to $\overline{AC}$ in Fig. 1-22, is perpendicular to $\overline{AC}$ and forms right angles 1 and 2. Each angle bisector, median, and altitude of a triangle extends from a vertex to the opposite side.

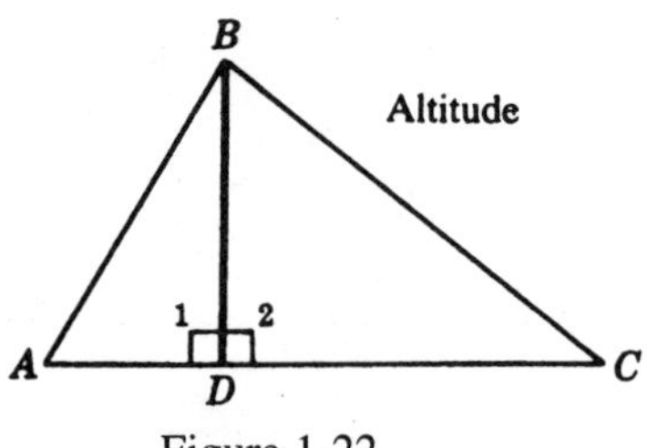

Figure 1-22

5. *Altitude of obtuse triangle*: In an obtuse triangle, the altitude drawn to either side of the obtuse angle falls outside the triangle. Thus, in obtuse triangle *ABC* (shaded) in Fig. 1-23, altitudes $\overline{BD}$ and $\overline{CE}$ fall outside the triangle. In each case, a side of the obtuse angle must be extended.

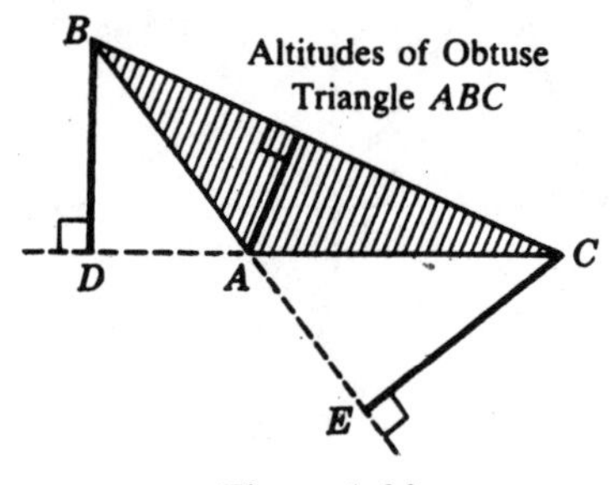

Figure 1-23

Pairs of Angles

Kinds of Pairs of Angles

1. *Adjacent angles*: Adjacent angles are two angles which have the same vertex and a common side between them. Thus the entire angle of $c°$ in Fig. 1-24 has been cut into two adjacent angles of $a°$ and $b°$. These adjacent angles have the same vertex A, and a common side $\overrightarrow{AD}$ between them. Here, $a° + b° = c°$.

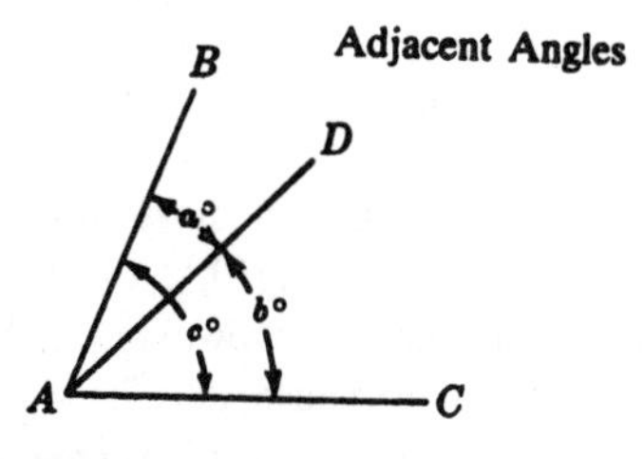

Figure 1-24

2. *Vertical angles*: Vertical angles are two nonadjacent angles formed by two intersecting lines. Thus, $\angle 1$ and $\angle 3$ in Fig. 1-25 are vertical angles formed by intersecting lines $\overleftrightarrow{AB}$ and $\overleftrightarrow{CD}$. Also, $\angle 2$ and $\angle 4$ are another pair of vertical angles formed by the same lines.

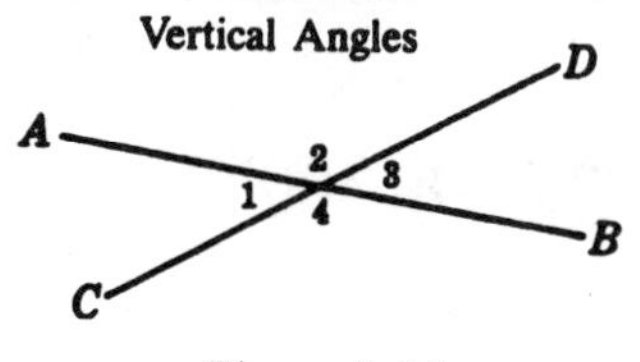

Figure 1-25

3. *Complementary angles*: Complementary angles are two angles whose measures total 90°. Thus, in Fig. 1-26(a), the angles of $a°$ and $b°$

are adjacent complementary angles. However, in (b) the complementary angles are nonadjacent. In each case, $a° + b° = 90°$. Either of two complementary angles is said to be the *complement* of the other.

Complementary Angles

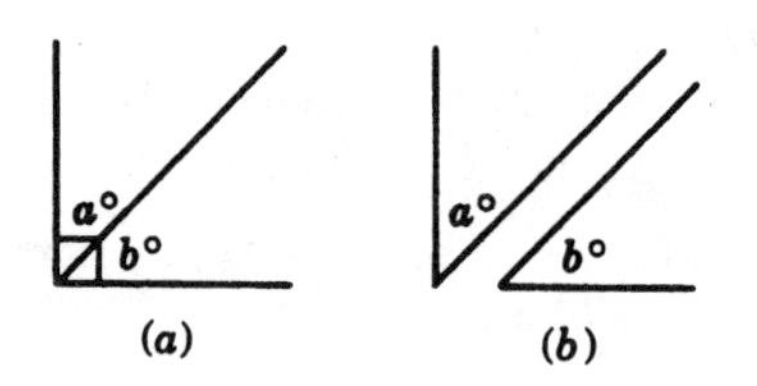

Figure 1-26

4. *Supplementary angles*: Supplementary angles are two angles whose measures total 180°. Thus, in Fig. 1-27(a), the angles of $a°$ and $b°$ are adjacent supplementary angles. However, in (b) the supplementary angles are nonadjacent. In each case, $a° + b° = 180°$. Either of two supplementary angles is said to be the *supplement* of the other.

Supplementary Angles

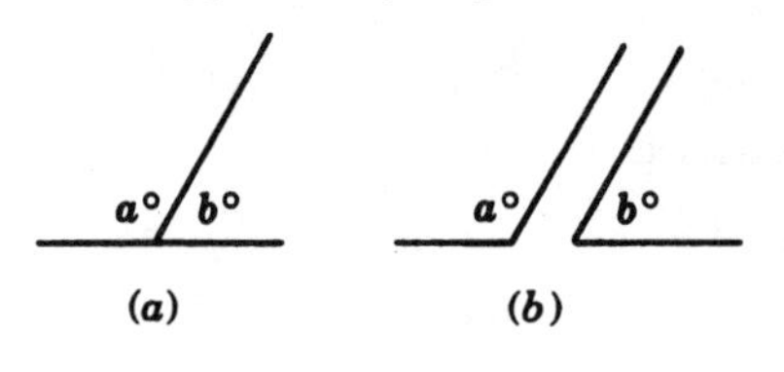

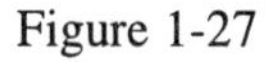
Figure 1-27

Principles of Pairs of Angles

Principle 1: *If an angle of* c° *is cut into two adjacent angles of* $a°$ *and* $b°$, *then* $a° + b° = c°$.

Thus, if $a° = 25°$ and $b° = 35°$ in Fig. 1-28, then $c° = 25° + 35° = 60°$.

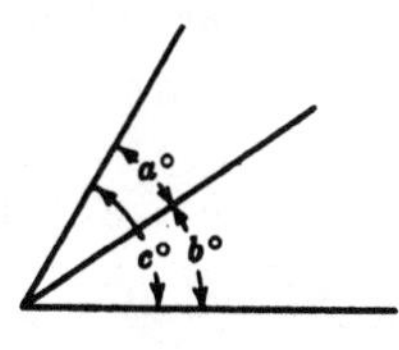

Figure 1-28

PRINCIPLE 2: *Vertical angles are congruent.*

Thus, if $\overleftrightarrow{AB}$ and $\overleftrightarrow{CD}$ are straight lines in Fig. 1-29, then $\angle 1 \cong \angle 3$ and $\angle 2 \cong \angle 4$. Hence, if $m\angle 1 = 40°$, then $m\angle 3 = 40°$; in such a case, $m\angle 2 = m\angle 4 = 140°$.

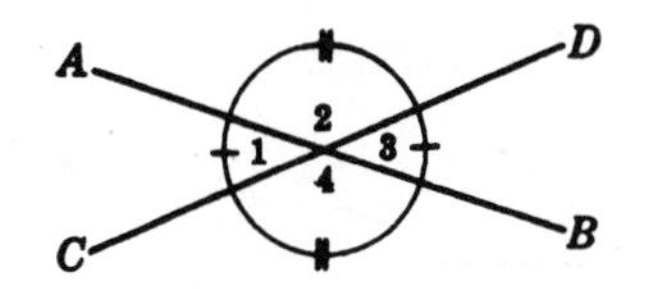

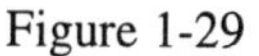

Figure 1-29

PRINCIPLE 3: *If two complementary angles contain $a°$ and $b°$, then $a° + b° = 90°$.*

Thus, if angles of $a°$ and $b°$ are complementary and $a° = 40°$, then $b° = 50°$[see Fig. 1-30].

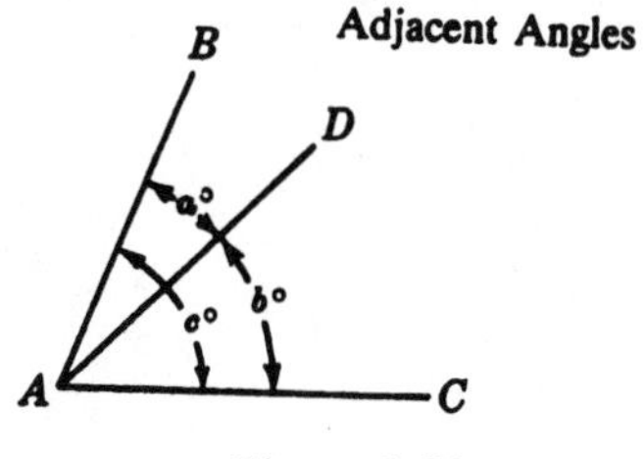

Figure 1-30

PRINCIPLE 4: *Adjacent angles are complementary if their exterior sides are perpendicular to each other.*

Thus, in Fig. 1-30, $a°$ and $b°$ are complementary angles since their exterior sides $\overline{AB}$ and $\overline{BC}$ are perpendicular to each other.

PRINCIPLE 5: *If two supplementary angles contain $a°$ and $b°$, then $a° + b° = 180°$.*

Thus, if angles of $a°$ and $b°$ are supplementary and $a° = 140°$, then $b° = 40°$ [see Fig. 1-31(a) or (b)].

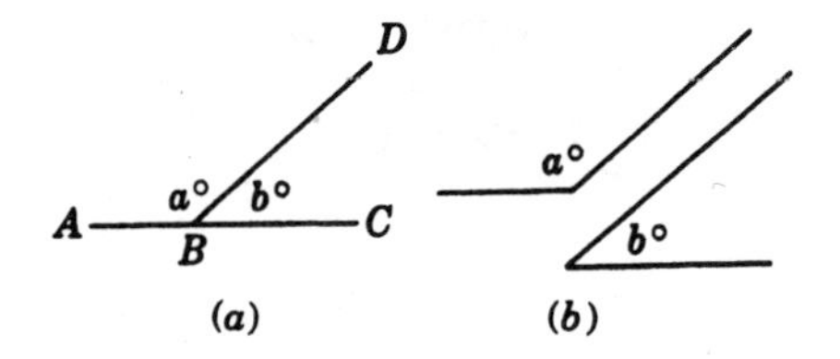

Figure 1-31

PRINCIPLE 6: *Adjacent angles are supplementary if their exterior sides lie in the same straight line.*

Thus, in Fig. 1-31(a), $a°$ and $b°$ are supplementary angles since their exterior sides $\overline{AB}$ and $\overline{BC}$ lie in the same straight line $\overleftrightarrow{AC}$.

PRINCIPLE 7: *If supplementary angles are congruent, each of them is a right angle. (Equal supplementary angles are right angles.)*

Thus, if $\angle 1$ and $\angle 2$ in Fig. 1-32 are both congruent and supplementary, then each of them is a right angle.

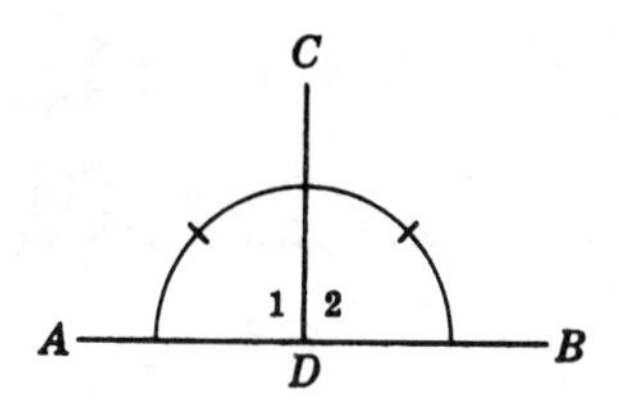

Figure 1-32

Chapter 2
DEDUCTIVE REASONING

IN THIS CHAPTER:

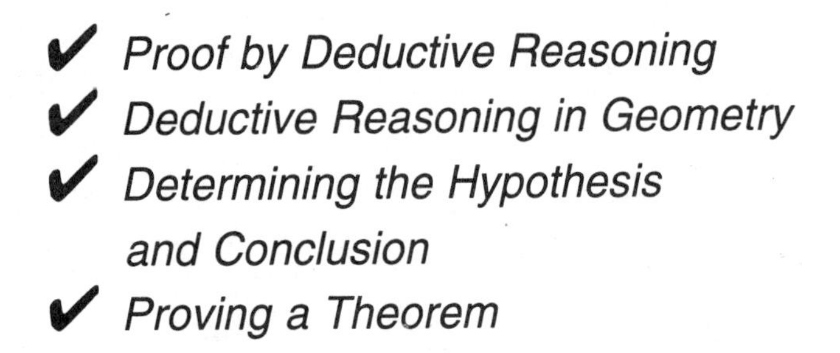

Proof by Deductive Reasoning

Deductive Reasoning is Proof

Deductive reasoning enables us to derive true or acceptably true conclusions from statements that are true or accepted as true. It consists of three steps as follows:

1. Making a *general statement* referring to a whole set or class of things, such as class of dogs: *All dogs are quadrupeds (have four feet).*

2. Making a *particular statement* about one or some of the members of the set or class referred to in the general statement: *All greyhounds are dogs.*

3. Making a *deduction* that follows logically when the general statement is applied to the particular statement: *All greyhounds are quadrupeds.*

Deductive reasoning is called *syllogistic reasoning* because the three statements together constitute a syllogism. In a syllogism, the general statement is called the *major premise*, the particular statement is the *minor premise*, and the deduction is the *conclusion*. Thus in the above syllogism:

1. The major premise is: *All dogs are quadrupeds.*

2. The minor premise is: *All greyhounds are dogs.*

3. The conclusion is: *All greyhounds are quadrupeds.*

Using a circle, as in Fig. 2-1, to represent each set or class will help you understand the relationships involved in deductive reasoning.

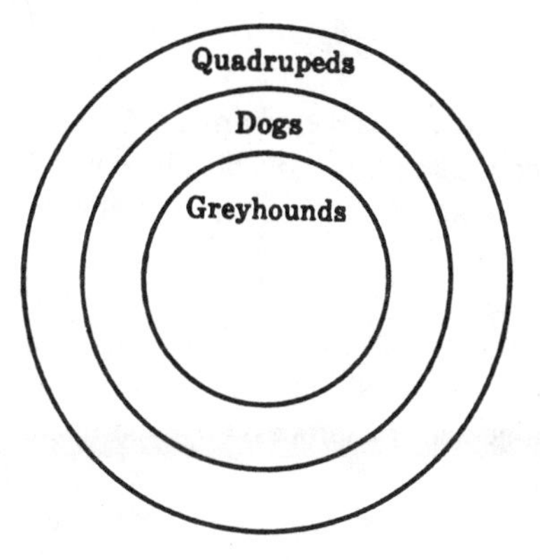

Figure 2-1

1. Since the major premise or general statement states that all dogs are quadrupeds, the circle representing dogs must be inside that for quadrupeds.

2. Since the minor premise or particular statement states that all greyhounds are dogs, the circle representing greyhounds are dogs must be inside that for dogs.

3. The conclusion is obvious. Since the circle of greyhounds must be inside the circle of quadrupeds, the only possible conclusion is that greyhounds are quadrupeds.

Observation, Measurement, and Experimentation Are Not Proof

1. *Observation cannot serve as proof.* Appearances may be misleading. Thus, in each part of Fig. 2-2, *AB* does not seem to equal *CD* although it actually does.

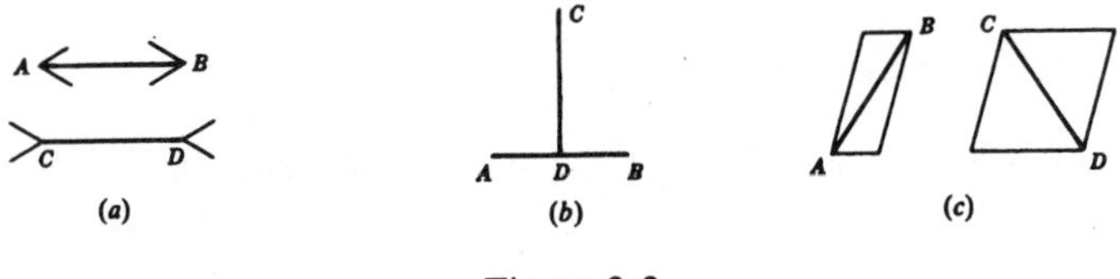

Figure 2-2

2. *Measurement cannot serve as proof.* Measurement applies only to the limited number of cases involved. The conclusion it provides is not exact but approximate, depending on the precision of the measuring instrument and the care of the observer.

3. *Experiment cannot serve as proof.* Its conclusions are only probable ones. The degree of probability depends on the particular situations or instances examined in the process of experimentation. Thus, it is probable that a pair of dice are loaded if ten successive 7s are rolled with the pair, and the probability is much greater if twenty successive 7s are rolled; however, neither probability is a certainty.

Deductive Reasoning in Geometry

The kinds of terms and statements discussed in this section comprise the deductive structure of geometry, which can be visualized as in Fig. 2-3.

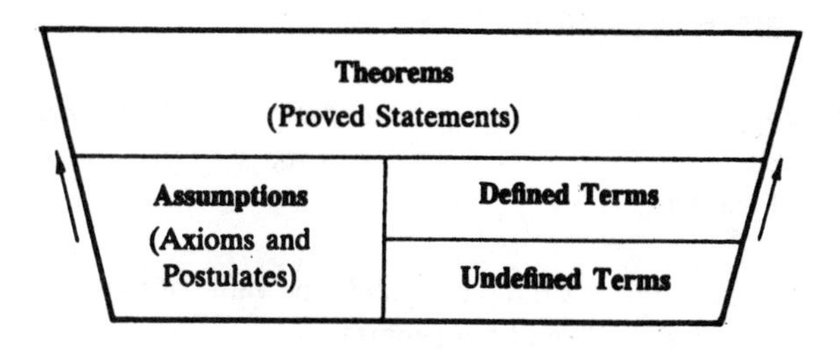

DEDUCTIVE STRUCTURE OF GEOMETRY

Figure 2-3

Undefined and Defined Terms

Point, line, and surface are the terms in geometry which are, by agreement, not defined. These undefined terms begin the process of definition in geometry and underlie the definitions of all other geometric terms.

We can define a triangle in terms of a polygon, a polygon in terms of a geometric figure, and a geometric figure as a figure composed of line segments, or parts of lines. However, the process of definition cannot be continued further because the term "line" is undefined.

Assumptions (Axioms and Postulates)

The entire structure of proof in geometry rests upon, or begins with, some unproved general statements called *postulates*. These are statements which we must willingly assume or accept as true so as to be able to deduce other statements.

When we draw a line segment between two points, we justify this by using as a reason the postulate "two points determine one and only one straight line." This reason is an assumption since we assume it to be true without requiring further justification.

Algebraic Postulates

Postulate 1: *Things equal to the same or equal things are equal to each other; if $a = b$ and $c = b$, then $a = c$.* (Transitive Postulate)

Thus, the total value of a dime is equal to the value of two nickels, since each is equal to the value of ten pennies.

Postulate 2: *A quantity may be substituted for its equal in any expression or equation.* (Substitution Postulate)

Thus, if $x = 5$ and $y = x + 3$, we may substitute 5 for x and find $y = 5 + 3 = 8$.

Postulate 3: *The whole equals the sum of its parts.* (Partition Postulate)

Thus, the total value of a dime, a nickel, and a penny is 16 cents.

Postulate 4: *Any quantity equals itself.* (Reflexive Postulate or Identity Postulate)

Thus, $x = x$, $m\angle A = m\angle A$, and $\overleftrightarrow{AB} = \overleftrightarrow{AB}$.

Postulate 5: *If equals are added to equals, the sums are equal; if $a = b$ and $c = d$, then $a + c = b + d$.* (Addition Postulate)

If	7 dimes = 70 cents	If	$x + y = 12$
and	2 dimes = 20 cents	and	$x - y = 8$
then	9 dimes = 90 cents	then	$2x = 20$

Postulate 6: *If equals are subtracted from equals, the differences are equal; if $a = b$ and $c = d$, then $a - c = b - d$.* (Subtraction Postulate)

If	7 dimes = 70 cents	If	$x + y = 12$
and	2 dimes = 20 cents	and	$x - y = 8$
then	5 dimes = 50 cents	then	$2y = 4$

Postulate 7: *If equals are multiplied by equals, the products are equal; if* $a = b$ *and* $c = d$, *then* $ac = bd$. (Multiplication Postulate)

Thus, if the price of one book is \$2, the price of three books is \$6.

Postulate 8: *If equals are divided by equals, the quotients are equal; if* $a = b$ *and* $c = d$, *then* $a/c = b/d$, *where* $c, d \neq 0$. (Division Postulate)

Thus, if the price of 4 lb of butter is 80 cents then, at the same rate, the price of 2 lb is 40 cents.

Postulate 9: *Like powers of equals are equal; if* $a = b$, *then* $a^n = b^n$. (Powers Postulate)

Thus, if $x = 5$, then $x^2 = 5^2$ or $x^2 = 25$.

Postulate 10: *Like roots of equals are equal; if* $a = b$, *then*

$$\sqrt[n]{a} = \sqrt[n]{b} .$$

Thus, if $y^3 = 27$, then $y = \sqrt[3]{27} = 3$.

Geometric Postulates

Postulate 11: *One and only one straight line can be drawn through any two points.*

Thus, $\overleftrightarrow{AB}$ is the only line that can be drawn between A and B in Fig. 2-4.

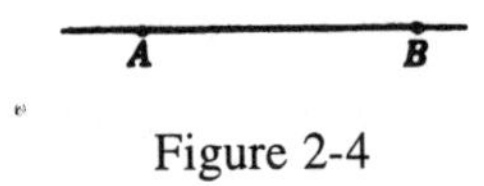

Figure 2-4

Postulate 12: *Two lines can intersect at one and only one point.*

Thus, only P is the point of intersection of $\overleftrightarrow{AB}$ and $\overleftrightarrow{CD}$ in Fig 2-5.

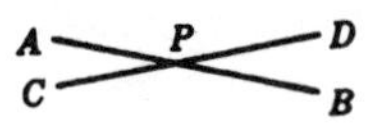

Figure 2-5

Postulate 13: *The length of a segment is the shortest distance between two points.*

Thus, $\overline{AB}$ is shorter than the curved or broken line segment between A and B in Fig 2-6.

Figure 2-6

Postulate 14: *One and only one circle can be drawn with any given point as center and a given line segment as a radius.*

Thus, only circle A in Fig. 2-7 can be drawn with A as the center and $\overline{AB}$ as a radius.

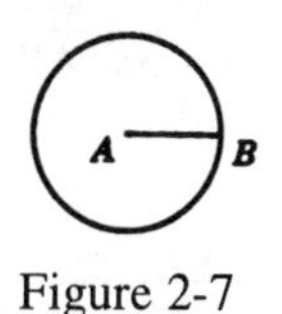

Figure 2-7

POSTULATE 15: *Any geometric figure can be moved without change in size or shape.*

Thus, ΔI in Fig. 2-8 can be moved to a new position without changing its size or shape.

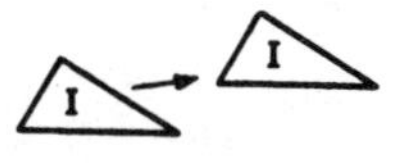

Figure 2-8

POSTULATE 16: *A segment has one and only one midpoint.*

Thus, only M is the midpoint of $\overline{AB}$ in Fig. 2-9.

Figure 2-9

POSTULATE 17: *An angle has one and only one bisector.*

Thus, only $\overrightarrow{AD}$ is the bisector of $\angle A$ in Fig. 2-10.

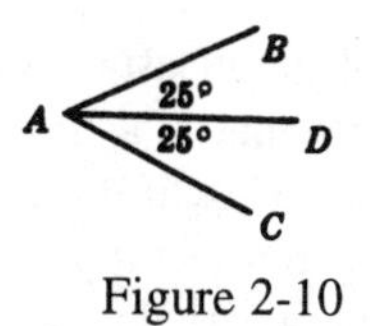

Figure 2-10

POSTULATE 18: *Through any point on a line, one and only one perpendicular can be drawn to the line.*

Thus, only $\overrightarrow{PC} \perp \overleftrightarrow{AB}$ at point P on $\overleftrightarrow{AB}$ (Fig. 2-11).

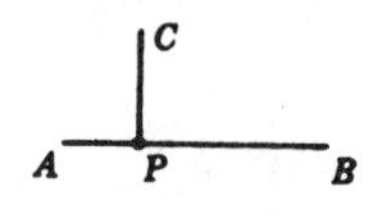

Figure 2-11

Postulate 19: *Through any point outside a line, one and only one perpendicular can be drawn to the given line.*

Thus, only $\overline{PC}$ can be drawn $\perp \overleftrightarrow{AB}$ from point P outside $\overleftrightarrow{AB}$ in Fig. 2-12.

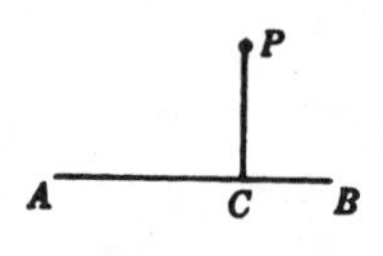

Figure 2-12

Theorems

Theorems are the statements that are proved in geometry. By using definitions and assumptions as reasons, we deduce or prove the basic theorems. As we use each new theorem to prove still more theorems, the process of deduction grows. However, if a new theorem is used to prove a previous one, the logical sequence is violated.

The theorem "the sum of the measures of the angles of a triangle equals 180°" is used to prove that "the sum of the measures of the angles of a pentagon is 540°." This, in turn, enables us to prove that "each angle of a regular pentagon measures 108°." However, it would be violating logical sequence if we tried to use the last theorem to prove either of the first two.

Principle 1: *All right angles are congruent.*

Thus, $\angle A \cong \angle B$ in Fig. 2-13.

Figure 2-13

Principle 2: *All straight angles are congruent.*

Thus, $\angle C \cong \angle D$ in Fig. 2-14.

Figure 2-14

Principle 3: *Complements of the same or of congruent angles are congruent.*

This is a combination of the following two principles:

1. *Complements of the same angle are congruent.* Thus, $\angle a \cong \angle b$ in Fig. 2-15; each is the complement of $\angle x$.

2. *Complements of congruent angles are congruent.* Thus, $\angle c \cong \angle d$ in Fig. 2-16; their complements are the congruent angles x and y.

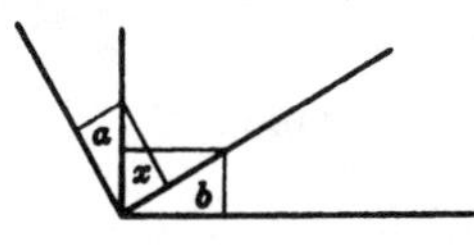

Figure 2-15

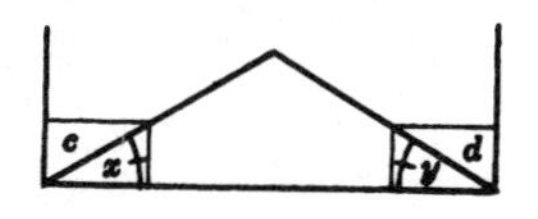

Figure 2-16

PRINCIPLE 4: *Supplements of the same or of congruent angles are congruent.*

This is a combination of the following two principles:

1. *Supplements of the same angle are congruent.* Thus, $\angle a \cong \angle b$ in Fig. 2-17; each is the supplement of $\angle x$.

2. *Supplements of congruent angles are congruent.* Thus, $\angle c \cong \angle d$ in Fig. 2-18; their supplements are the congruent angles x and y.

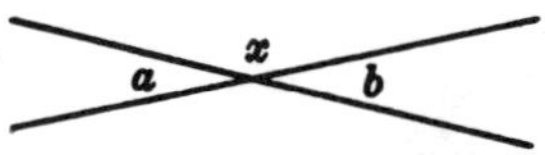

Figure 2-17

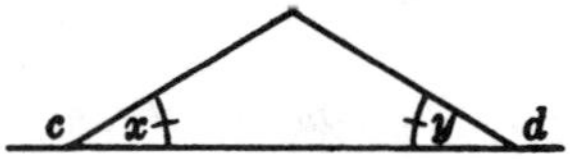

Figure 2-18

PRINCIPLE 5: *Vertical angles are congruent.*

Thus, in Fig. 2-19, $\angle a \cong \angle b$; this follows from Principle 4, since $\angle a$ and $\angle b$ are supplements of the same angle, $\angle c$.

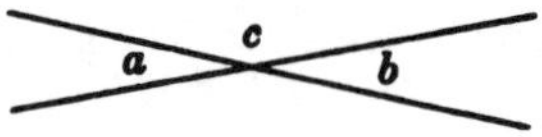

Figure 2-19

Determining the Hypothesis and Conclusion

Statement Forms: Subject-Predicate Form and If-Then Form

The statements "a heated metal expands" and "if a metal is heated, then it expands" are two forms of the same idea. The following table shows how each form may be divided into its two important parts, the *hypothesis*, which tells *what is given*, and the *conclusion*, which tells *what is to be proved*. Note that, in the if-then form, the word *then* may be omitted.

Form	Hypothesis (What is given)	Conclusion (What is to be proved)
Subject-predicate form: *A heated metal expands.*	**Hypothesis is subject:** *A heated metal*	**Conclusion is predicate:** *expands*
If-then form: *If a metal is heated, then it expands.*	**Hypothesis is if clause:** *If a metal is heated*	**Conclusion is then clause:** *then it expands*

Converse of a Statement

The *converse of a statement* is formed by interchanging the hypothesis and conclusion. Hence to form the converse of an if-then statement, interchange the if and then clauses. In the case of the subject-predicate form, interchange the subject and predicate.

Thus, the converse of "triangles are polygons" is "polygons are triangles." Also, the converse of "if a metal is heated, then it expands" is "if a metal expands, then it is heated." Note in each of these cases that the statement is true but its converse need not necessarily be true.

PRINCIPLE 1: *The converse of a true statement is not necessarily true.*

Thus, the statement "triangles are polygons" is true. Its converse need not be true.

Principle 2: *The converse of a definition is always true.*

Thus, the converse of the definition "a triangle is a polygon of three sides" is "a polygon of three sides is a triangle." Both the definition and its converse are true.

Solved Problem 2-1. Determine the hypothesis and conclusion of each statement.

Statements	Solutions	
	Hypothesis (subject)	**Conclusion** (predicate)
(*a*) Perpendiculars form right angles.	Perpendiculars	form right angles
(*b*) Complements of the same angle are congruent.	Complements of the same angle	are congruent
(*c*) An equilateral triangle is equiangular.	An equilateral triangle	is equiangular
(*d*) A right triangle has only one right angle.	A right triangle	has only one right angle
(*e*) A triangle is not a quadrilateral.	A triangle	is not a quadrilateral

Solved Problem 2-2. Determine the hypothesis and conclusion of each statement.

Statements	Solutions	
	Hypothesis (if clause)	**Conclusion** (then clause)
(*a*) If a line bisects an angle, then it divides the angle into two congruent parts.	If a line bisects an angle	then it divides the angle into two congruent parts
(*b*) A triangle has an obtuse angle if it is an obtuse triangle.	If it is an obtuse triangle	(then) a triangle has an obtuse angle
(*c*) If a student is sick, she should not go to school.	If a student is sick	(then) she should not go to school
(*d*) A student, if he wishes to pass, must study regularly.	If he wishes to pass	(then) a student must study regularly

Proving a Theorem

Theorems should be proved using the following step-by-step procedure. The form of the proof is shown in the example that follows the procedure.

1. Divide the theorem into its hypothesis (what is given) and its conclusion (what is to be proved). Underline the hypothesis with a single line, and the conclusion with a double line.

2. On one side, make a marked diagram. Markings on the diagram should include such helpful symbols as square corners for right angles, cross marks for equal parts, and question marks for parts to be proved equal.

3. On the other side, next to the diagram, state what is given and what is to be proved. The "Given" and "To Prove" must refer to the parts of the diagram.

4. Present a plan. Although not essential, a plan is very advisable. It should state the major methods of proof to be used.

5. On the left, present statements in successively numbered steps. The last statement must be the one to be proved. All the statements must refer to parts of the diagram.

6. On the right, next to the statements, provide a reason for each statement. Acceptable reasons in the proof of a theorem are given facts, definitions, postulates, assumed theorems, and previously proven theorems.

Step 1: **Prove:** All right angles are equal in measure.

Steps 2 and 3: **Given:** $\angle A$ and $\angle B$ are rt. $\angle s$
To Prove: $m\angle A = m\angle B$

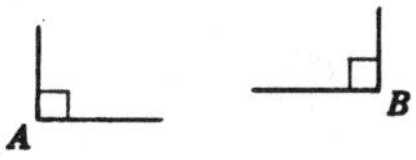

Step 4: **Plan:** Since each angle equals 90°, the angles are equal in measure, using Post. 1: Things equal to the same thing are equal to each other.

A B

Steps 5 and 6:

Statements	Reasons
1. $\angle A$ and $\angle B$ are rt. $\angle$s.	1. Given
2. $m\angle A$ and $m\angle B$ each = 90°.	2. m(rt. $\angle$) = 90°
3. $m\angle A = m\angle B$	3. Things = to same thing = each other.

Solved Problem 2-3. Use the proof procedure to prove that supplements of angles of equal measure have equal measure.

Step 1: **Prove:** Supplements of angles of equal measure have equal measure.

Steps 2 and 3: **Given:** $\angle a$ sup. $\angle 1$, $\angle b$ sup. $\angle 2$
$m\angle 1 = m\angle 2$

To Prove: $m\angle a = m\angle b$

Step 4: **Plan:** Using the subtraction postulate, the equal angle measures may be subtracted from the equal sums of measures of pairs of supplementary angles. The equal remainders are the measures of the supplements.

Steps 5 and 6:

Statements	Reasons
1. $\angle a$ sup. $\angle 1$, $\angle b$ sup. $\angle 2$	1. Given
2. $m\angle a + m\angle 1 = 180°$ $m\angle b + m\angle 2 = 180°$	2. Sup. $\angle$s are $\angle$s the sum of whose measures = 180°.
3. $m\angle a + m\angle 1 = m\angle b + m\angle 2$	3. Things = to the same thing = each other.
4. $m\angle 1 = m\angle 2$	4. Given
5. $m\angle a = m\angle b$	5. If =s are subtracted from =s, the differences are =.

Chapter 3
Congruent Triangles

In This Chapter:

- ✔ *Congruent Triangles*
- ✔ *Isosceles and Equilateral Triangles*

Congruent Triangles

Congruent figures are figures that have the same size and the same shape; they are the exact duplicates of each other. Such figures can be made to coincide so that their corresponding parts will fit together. Two circles having the same radius are congruent circles.

Congruent triangles are triangles that have the same size and the same shape.

If two triangles are congruent, their corresponding sides and angles must be congruent. Thus, congruent triangles ABC and $A'B'C'$ in Fig. 3-1 have congruent corresponding sides ($AB \cong A'B'$, $BC \cong B'C'$, and $AC \cong A'C'$) and congruent corresponding angles ($\angle A \cong \angle A'$, $\angle B \cong \angle B'$, $\angle C \cong \angle C'$).

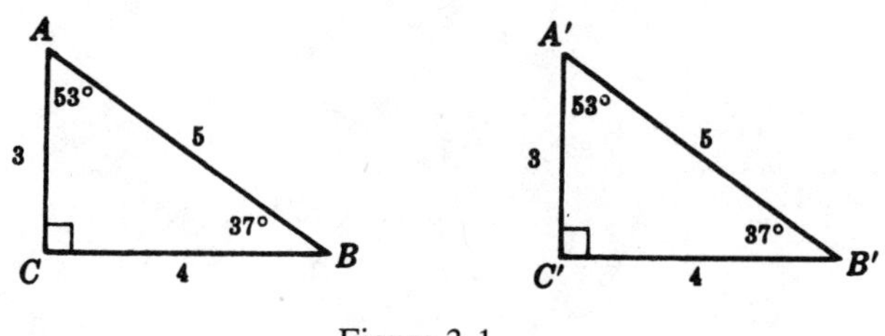

Figure 3-1

Basic Principles of Congruent Triangles

PRINCIPLE 1: *If two triangles are congruent, then their corresponding parts are congruent.* (Corresponding parts of congruent triangles are congruent.)

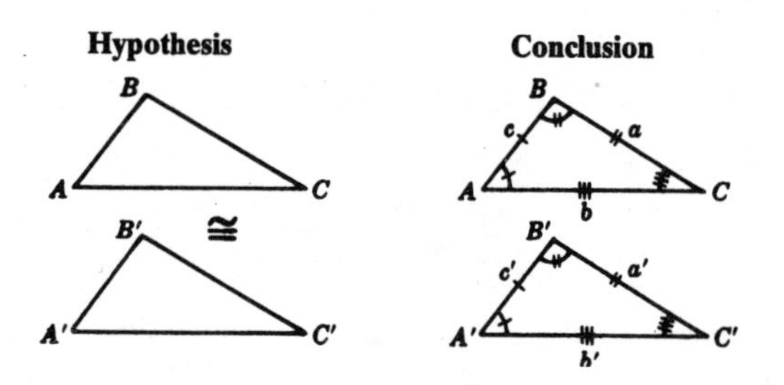

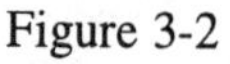

Figure 3-2

PRINCIPLE 2: (s.a.s. ≅ s.a.s.) *If two sides and the included angle of one triangle are congruent to the corresponding parts of another, then the triangles are congruent.*

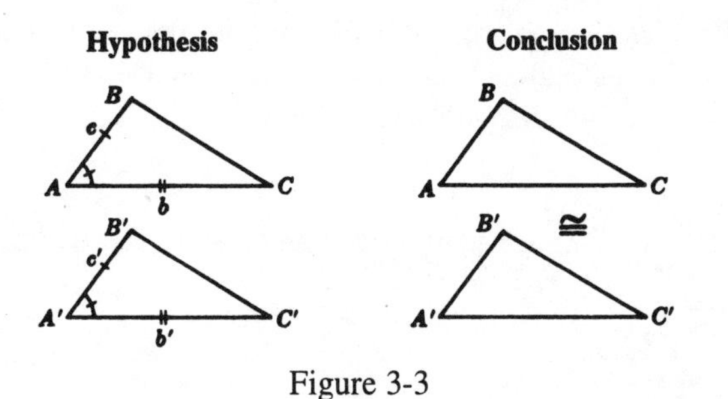

Figure 3-3

Principle 3: (a.s.a. ≅ a.s.a.) *If two angles and the included side of one triangle are congruent to the corresponding parts of another, then the triangles are congruent.*

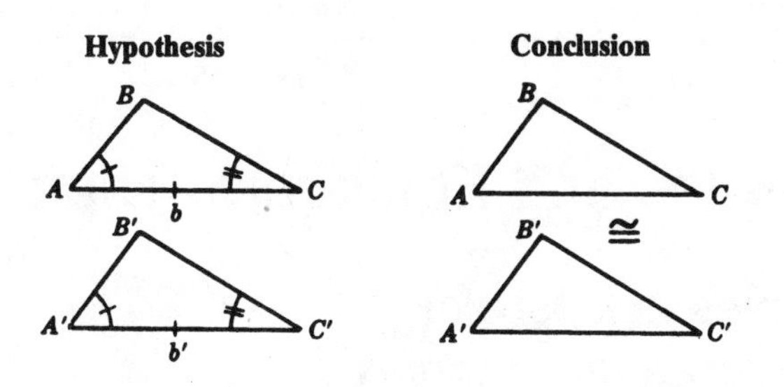

Figure 3-4

Principle 4: (s.s.s. ≅ s.s.s.) *If three sides of one triangle are congruent to three sides of another, then the triangles are congruent.*

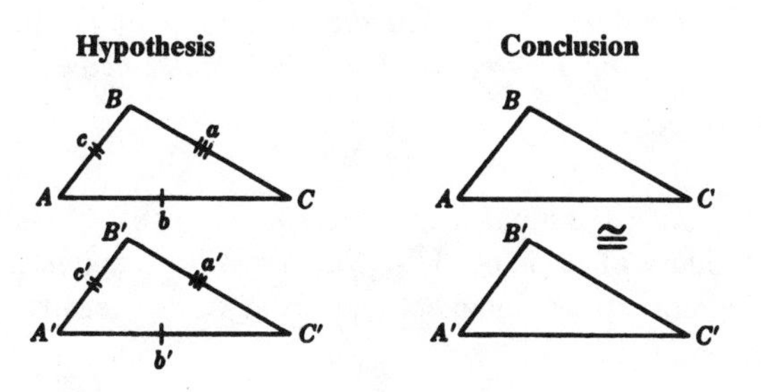

Figure 3-5

Solved Problem 3-1. Prove that if the opposite sides of a quadrilateral are equal and a diagonal is drawn, equal angles are formed between the diagonal and the sides.

Solution

If the opposite sides of a quadrilateral are congruent and a diagonal is drawn, congruent angles are formed between the diagonal and the sides.

Given: Quadrilateral $ABCD$
$\overline{AB} \cong \overline{CD}$, $\overline{BC} \cong \overline{AD}$
$\overline{AC}$ is a diagonal.

To Prove: $\angle 1 \cong \angle 4$, $\angle 2 \cong \angle 3$

Plan: Prove $\Delta I \cong \Delta II$

PROOF:

Statements	Reasons
1. $\overline{AB} \cong \overline{CD}$, $\overline{BC} \cong \overline{AD}$	1. Given
2. $\overline{AC} \cong \overline{AC}$	2. Reflexive property
3. $\Delta I \cong \Delta II$	3. s.s.s. ≅ s.s.s.
4. $\angle 1 \cong \angle 4$, $\angle 2 \cong \angle 3$	4. Corresponding parts of ≅ ▲are ≅.

Isosceles and Equilateral Triangles

Principles of Isosceles and Equilateral Triangles

Principle 1: *If two sides of a triangle are congruent, the angles opposite these sides are congruent.* (Base angles of an isosceles triangle are congruent.)

Principle 2: *If two angles of a triangle are congruent, the sides opposite these angles are congruent.* (Principle 2 is the converse of Principle 1.)

Principle 3: *An equilateral triangle is equiangular.* (Principle 3 is a corollary of Principle 1. A *corollary* of a theorem is another theorem whose statement and proof follow readily from the theorem.)

PRINCIPLE 4: *An equiangular triangle is equilateral.* (Principle 4 is the converse of Principle 3 and a corollary of Principle 2.)

Solved Problem 3-2. Prove that the bisector of the vertex angle of an isosceles triangle is a median to the base.

Solution

The bisector of the vertex angle of an isosceles triangle is a median to the base.

Given: Isosceles $\triangle ABC$ ($\overline{AB} \cong \overline{BC}$)
$\overline{BD}$ bisects $\angle B$

To Prove: $\overline{BD}$ is a median to $\overline{AC}$

Plan: Prove $\triangle I \cong \triangle II$ to obtain $\overline{AD} \cong \overline{DC}$.

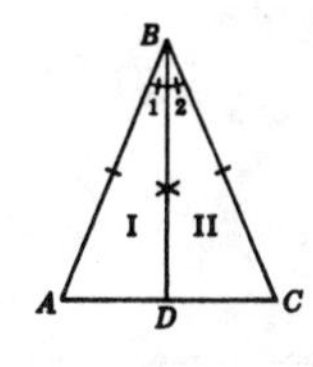

PROOF:

Statements	Reasons
1. $\overline{AB} \cong \overline{BC}$	1. Given
2. $\overline{BD}$ bisects $\angle B$.	2. Given
3. $\angle 1 \cong \angle 2$	3. To bisect is to divide into two congruent parts.
4. $\overline{BD} \cong \overline{BD}$	4. Reflexive property
5. $\triangle I \cong \triangle II$	5. s.a.s. ≅ s.a.s.
6. $\overline{AD} \cong \overline{DC}$	6. Corresponding parts of ≅ ▲ are ≅.
7. $\overline{BD}$ is a median to $\overline{AC}$.	7. A line from a vertex of a $\triangle$ bisecting opposite side is a median.

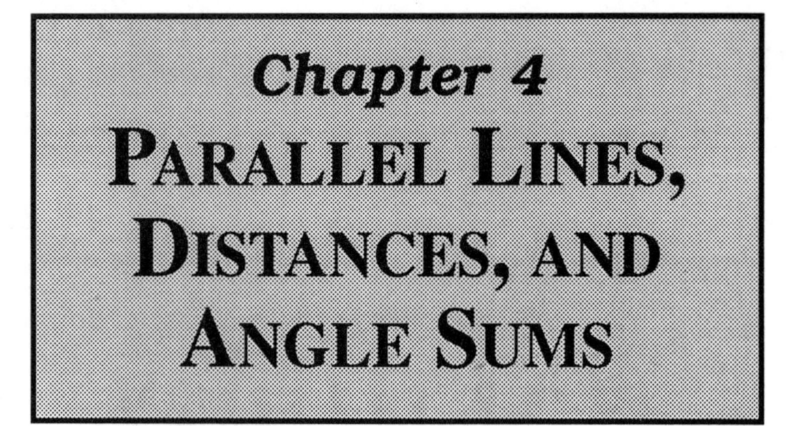

IN THIS CHAPTER:

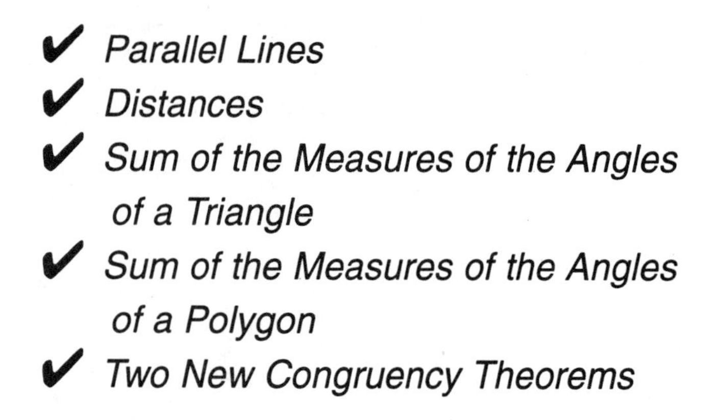

Parallel Lines

Parallel lines are straight lines that lie in the same plane and do not intersect however far they are extended. The symbol for parallel is $\parallel$; thus $\overleftrightarrow{AB} \parallel \overleftrightarrow{CD}$ is read "line $\overleftrightarrow{AB}$ is parallel to line $\overleftrightarrow{CD}$." (See Fig. 4-1.)

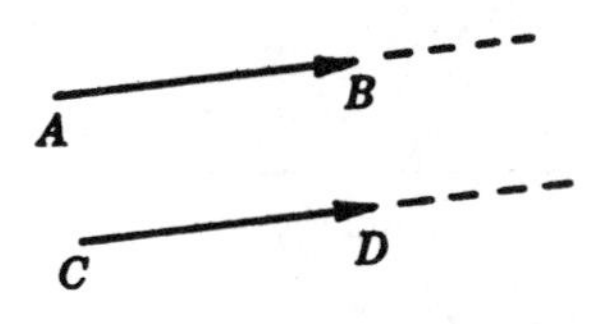

Figure 4-1

A *transversal* of two or more lines is a line that cuts across these lines. Thus $\overleftrightarrow{EF}$ is a transversal of $\overleftrightarrow{AB}$ and $\overleftrightarrow{CD}$, in Fig. 4-2.

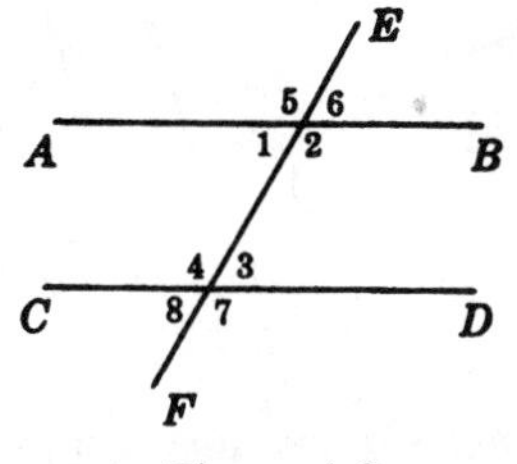

Figure 4-2

The *interior angles* formed by two lines cut by a transversal are the angles between the two lines, while the *exterior angles* are those outside the lines. Thus, of the eight angles formed by $\overleftrightarrow{AB}$ and $\overleftrightarrow{CD}$ cut by $\overleftrightarrow{EF}$ in Fig. 4-2, the interior angles are $\angle 1$, $\angle 2$, $\angle 3$, and $\angle 4$; the exterior angles are $\angle 5$, $\angle 6$, $\angle 7$, and $\angle 8$.

Pairs of Angles Formed by Two Lines Cut by a Transversal

Corresponding angles of two lines cut by a transversal are angles on the same side of the transversal and on the same side of the lines. Thus, $\angle 1$ and $\angle 2$ in Fig. 4-3 are corresponding angles of $\overleftrightarrow{AB}$ and $\overleftrightarrow{CD}$ cut by transversal $\overleftrightarrow{EF}$. When two parallel lines are cut by a transversal, the sides of two corresponding angles form a capital F in varying positions, as shown in Fig. 4-4.

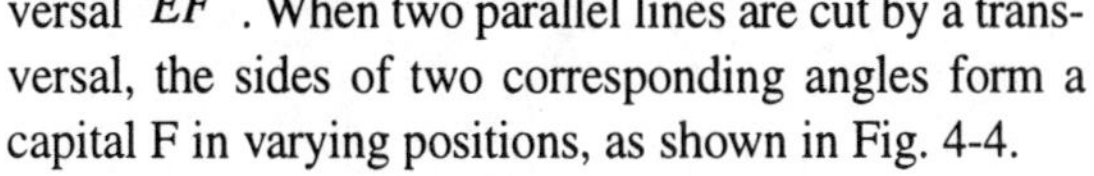

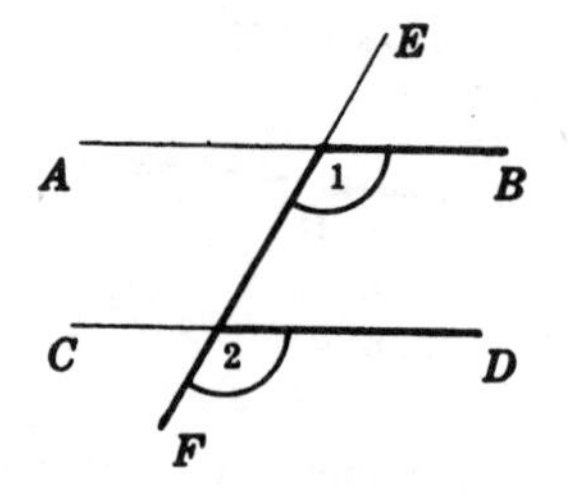

Figure 4-3

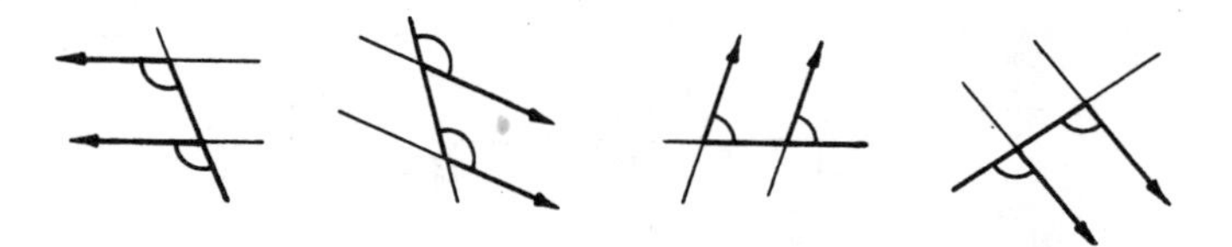

Figure 4-4

Alternate interior angles of two lines cut by a transversal are non-adjacent angles between the two lines and on opposite sides of the transversal. Thus, $\angle 1$ and $\angle 2$ in Fig. 4-5 are alternate interior angles of $\overleftrightarrow{AB}$ and $\overleftrightarrow{CD}$ cut by $\overleftrightarrow{EF}$. When parallel lines are cut by a transversal, the sides of two alternate interior angles form a capital Z or N in varying positions, as shown in Fig. 4-6.

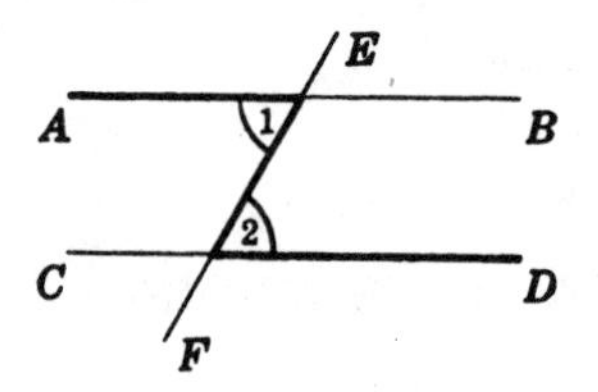

Figure 4-5

Figure 4-6

When parallel lines are cut by a transversal, *interior angles on the same side of the transversal* can be readily located by noting the capital U formed by their sides (Fig. 4-7).

Figure 4-7

Principles of Parallel Lines

PRINCIPLE 1: *Through a given point not on a given line, one and only one line can be drawn parallel to a given line.* (Parallel Line Postulate)

PRINCIPLE 2: *Two lines are parallel if a pair of corresponding angles are congruent.*

PRINCIPLE 3: *Two lines are parallel if a pair of alternate interior angles are congruent.*

PRINCIPLE 4: *Two lines are parallel if a pair of interior angles on the same side of a transversal are supplementary.*

PRINCIPLE 5: *Lines are parallel if they are perpendicular to the same line. (Perpendiculars to the same line are parallel.)*

PRINCIPLE 6: *Lines are parallel if they are parallel to the same line. (Parallels to the same line are parallel.)*

PRINCIPLE 7: *If two lines are parallel, each pair of corresponding angles are congruent. (Corresponding angles of parallel lines are congruent.)*

PRINCIPLE 8: *If two lines are parallel, each pair of alternate interior angles are congruent. (Alternate interior angles of parallel lines are congruent.)*

PRINCIPLE 9: *If two lines are parallel, each pair of interior angles on the same side of the transversal are supplementary.*

PRINCIPLE 10: *If lines are parallel, a line perpendicular to one of them is perpendicular to others also.*

PRINCIPLE 11: *If lines are parallel, a line parallel to one of them is parallel to the others also.*

PRINCIPLE 12: *If the sides of two angles are respectively parallel to each other, the angles are either congruent or supplementary.*

Distances

Distances Between Two Geometric Figures

The distance between two geometric figures is the straight line segment that is the *shortest segment between the figures.*

1. The distance *between two points*, such as P and Q in Fig. 4-8(a), is the line segment $\overline{PQ}$ between them.

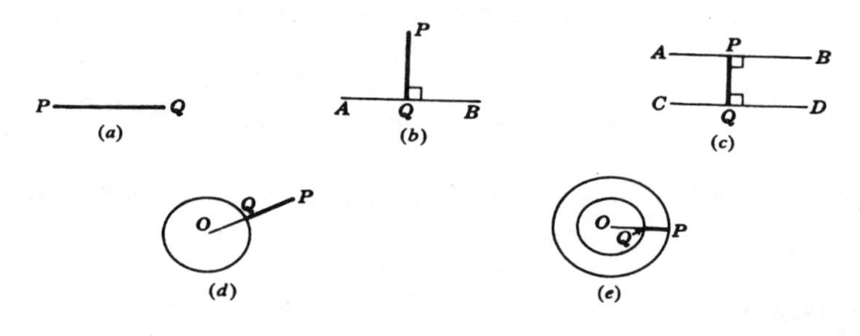

Figure 4-8

2. The distance *between a point and a line*, such as P and $\overleftrightarrow{AB}$ in Fig. 4-8(b), is the line segment $\overline{PQ}$, the perpendicular from the point to the line.

3. The distance *between two parallels*, such as $\overleftrightarrow{AB}$ and $\overleftrightarrow{CD}$ in Fig. 4-8(c), is the line segment $\overline{PQ}$, a perpendicular between the two parallels.

4. The distance *between a point and a circle*, such as P and the circle in Fig. 4-8(d), is the line segment $\overline{PQ}$, the segment of $\overline{OP}$ between the point and the circle.

5. The distance *between two concentric circles*, such as two circles whose center is O, is the line segment $\overline{PQ}$, the segment of the larger radius that lies between the two circles, as shown in Fig. 4-8(e).

Distance Principles

PRINCIPLE 1: *If a point is on the perpendicular bisector of a line segment, then it is equidistant from the ends of a line segment.*

PRINCIPLE 2: *If a point is equidistant from the ends of a line segment, then it is on the perpendicular bisector of the line segment.* (Principle 2 is the converse of Principle 1.)

PRINCIPLE 3: *If a point is on the bisector of an angle, then it is equidistant from the sides of the angle.*

PRINCIPLE 4: *If a point is equidistant from the sides of an angle, then it is on the bisector of the angle.* (Principle 4 is the converse of Principle 3.)

PRINCIPLE 5: *Two points each equidistant from the ends of a line segment determine the perpendicular bisector of the line segment.* (The line joining the vertices of two isosceles triangles having a common base is the perpendicular bisector of the base.)

PRINCIPLE 6: *The perpendicular bisectors of the sides of a triangle meet in a point which is equidistant from the vertices of the triangle.*

PRINCIPLE 7: *The bisectors of the angles of a triangle meet in a point which is equidistant from the sides of the triangle.*

Sum of the Measures of the Angles of a Triangle

The angles of any triangle may be torn off, as in Fig. 4-9(a), and then fitted together, as shown in (b). The three angles will form a straight angle.

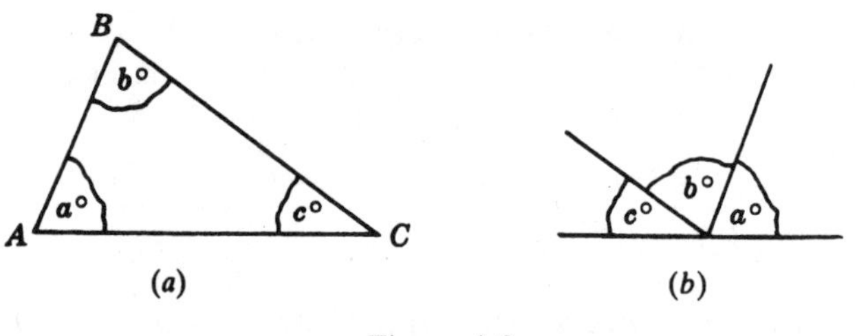

Figure 4-9

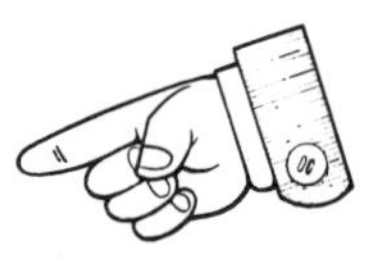

We can prove that the sum of the measures of the angles of a triangle equals 180° by drawing a line through one vertex of the triangle parallel to the side opposite the vertex. In Fig. 4-9, $\overleftrightarrow{MN}$ is drawn through B parallel to AC. Note that the measure of the straight angle at B equals the sum of the measures of the angles of $\triangle ABC$; that is, $a° + b° + c° = 180°$. Each pair of congruent angles is a pair of alternate interior angles of parallel lines.

Interior and Exterior Angles of a Polygon

An exterior angle of a polygon is formed whenever one of its sides is extended through a vertex. If each of the sides of a polygon is extended, as shown in Fig. 4-10, an exterior angle will be formed at each vertex. Each of these exterior angles is the supplement of its adjacent interior angle.

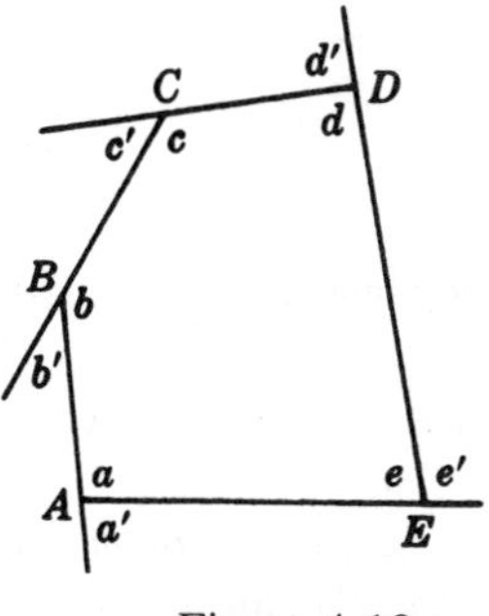

Figure 4-10

Thus, in each case of pentagon *ABCDE*, there will be five exterior angles, one at each vertex. Note that each exterior angle is the supplement of an adjacent interior angle. For example, $m\angle a + m\angle a' = 180°$.

Angle-Measure-Sum Principles

PRINCIPLE 1: *The sum of the measures of the angles of a triangle equals the measure of a straight angle.*

PRINCIPLE 2: *If two angles of one triangle are congruent respectively to two angles of another triangle, the remaining angles are congruent.*

PRINCIPLE 3: *The sum of the measures of the angles of a quadrilateral equals* 360°.

PRINCIPLE 4: *The measure of each exterior angle of a triangle equals the sum of the measures of its two nonadjacent interior angles.*

PRINCIPLE 5: *The sum of the measures of the exterior angles of a triangle equals* 360°.

PRINCIPLE 6: *The measure of each angle of an equilateral triangle equals* 60°.

Principle 7: *The acute angles of a right triangle are complementary.*

Principle 8: *The measure of each acute angle of an isosceles right triangle equals* 45°.

Principle 9: *A triangle can have no more than one right angle.*

Principle 10: *A triangle can have no more than one obtuse angle.*

Principle 11: *Two angles are congruent or supplementary if their sides are respectively perpendicular to each other.*

Sum of the Measures of the Angles of a Polygon

A *polygon* is a closed plane figure bounded by straight line segments as sides. An *n-gon* is a polygon of n sides. Thus, a polygon of 20 sides is a 20-gon.

Names of Polygons According to the Number of Sides

Number of Sides	Polygon	Number of Sides	Polygon
3	Triangle	8	Octagon
4	Quadrilateral	9	Nonagon
5	Pentagon	10	Decagon
6	Hexagon	12	Dodecagon
7	Heptagon	n	n-gon

A *regular polygon* is an equilateral and equiangular polygon. Thus, a regular pentagon is a polygon having 5 congruent angles and 5 congruent sides (Fig. 4-11). A square is a regular polygon of 4 sides.

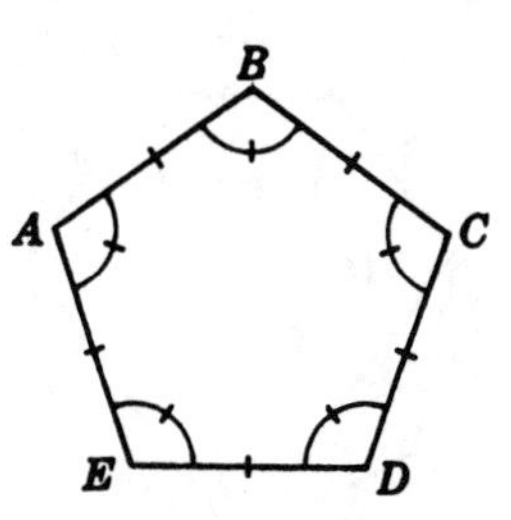

Regular Pentagon

Figure 4-11

Sum of the Measures of the Interior Angles of a Polygon

By drawing diagonals from any vertex to each of the other vertices, as in Fig. 4-12, a polygon of 7 sides is divisible into 5 triangles. Note that each triangle has one side of the polygon, except the first and last triangles which have two such sides.

In general, this process will divide a polygon of n sides into $n - 2$ triangles; that is, the number of such triangles is always two less than the number of sides of the polygon.

The sum of the measures of the interior angles of the polygon equals the sum of the measures of the interior angles of the triangles. Hence:

Sum of measures of interior angles of a polygon of n sides $= (n - 2)180°$

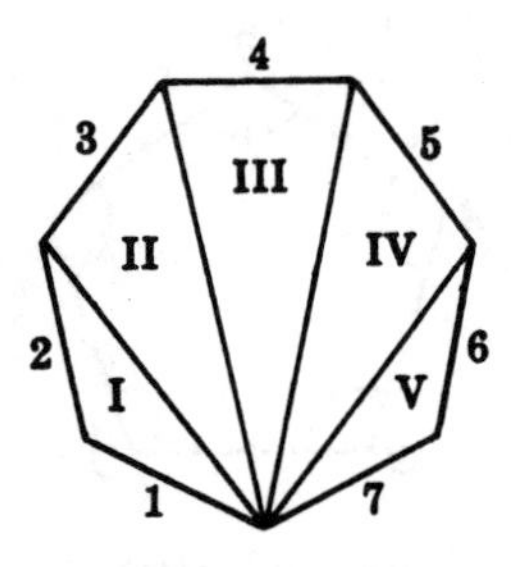

Figure 4-12

Sum of the Measures of the Exterior Angles of a Polygon

The exterior angles of a polygon can be reproduced together, so that they have the same vertex. To do this, draw lines parallel to the sides of the polygon from a point, as shown in Fig. 4-13. If this is done, it can be seen that regardless of the number of sides, the sum of the measures of the exterior angles equals 360°. Hence:

Sum of measures of exterior angles of a polygon of n sides = 360°

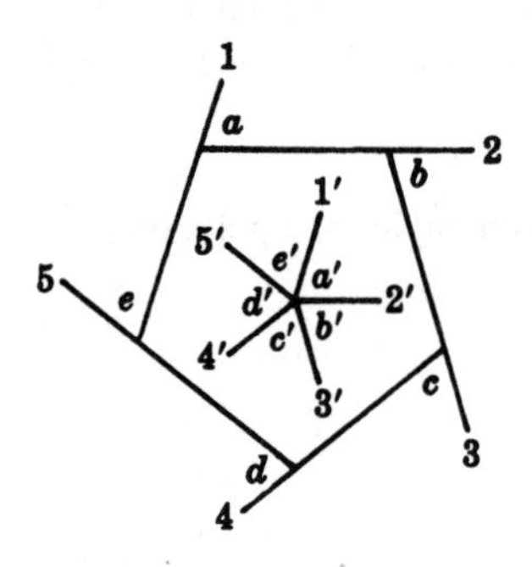

Figure 4-13

Polygon-Angle Principles

For any polygon

PRINCIPLE 1: *If S is the sum of the measures of the interior angles of a polygon of n sides, then*

$$S = n - 2 \text{ straight angles} = (n - 2)180°$$

PRINCIPLE 2: *The sum of the measures of the exterior angles of any polygon equals* 360°.

For a regular polygon

PRINCIPLE 3: *If a regular polygon of n sides (Fig. 4-14) has an interior angle of measure i and an exterior angle of measure e (in degrees), then*

$$i = \frac{180(n-2)}{n} \quad e = \frac{360}{n} \quad \text{and } i + e = 180$$

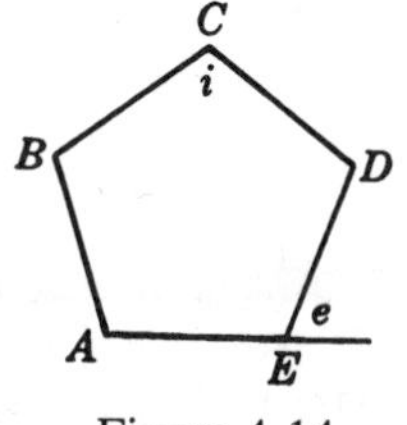

Figure 4-14

Thus, for a regular polygon of 20 sides,

$$i = \frac{180(20-2)}{20} = 162$$

$$e = \frac{360}{20} = 18$$

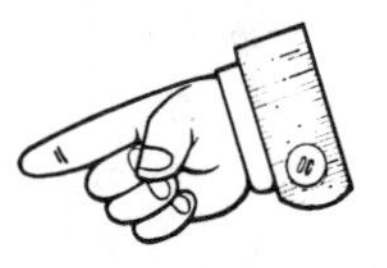

$$i + e = 162 + 18 = 180$$

Two New Congruency Theorems

Three methods of proving triangles congruent have already been introduced here. These are:

1. s.a.s. $\cong$ s.a.s.

2. a.s.a. $\cong$ a.s.a.

3. s.s.s. $\cong$ s.s.s.

Two additional methods of proving that triangles are congruent are:

4. s.a.a. $\cong$ s.a.a.

5. hy. leg $\cong$ hy. leg

Two New Congruency Principles

PRINCIPLE 1: (s.a.a. $\cong$ s.a.a.) *If two angles and a side opposite one of them of one triangle are congruent to the corresponding parts of another, the triangles are congruent.*

PRINCIPLE 2: (hy. leg $\cong$ hy. leg) *If the hypotenuse and a leg of one right triangle are congruent to the corresponding parts of another right triangle, the triangles are congruent.*

Solved Problem 4-1. (a) Prove that if the measure of one angle of a triangle equals the sum of the measures of the other two, then the triangle is a right triangle. (b) Prove that if the opposite angles of a quadrilateral are congruent, then its opposite sides are parallel.

Solutions

(*a*) **Given:** $\triangle ABC$, $m\angle C = m\angle A + m\angle B$

To Prove: $\triangle ABC$ is a right triangle.

Plan: Prove $m\angle C = 90°$

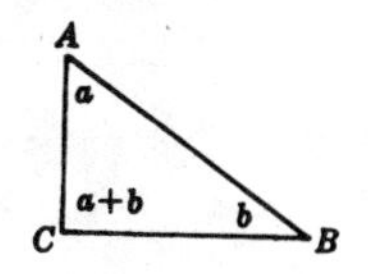

ALGEBRAIC PROOF:

Let a = number of degrees in $\angle A$

b = number of degrees in $\angle B$

Then $a + b$ = number of degrees in $\angle C$

$$a + b + (a + b) = 180 \quad \text{(Pr. 1)}$$
$$2a + 2b = 180$$
$$a + b = 90$$

Since $m\angle C = 90°$, $\triangle ABC$ is a rt. $\triangle$.

(*b*) **Given:** Quadrilateral $ABCD$

$\angle A \cong \angle C$, $\angle B \cong \angle D$

To Prove: $\overline{AB} \parallel \overline{CD}$, $\overline{BC} \parallel \overline{AD}$

Plan: Prove int. ∡ on same side of transversal are supplementary.

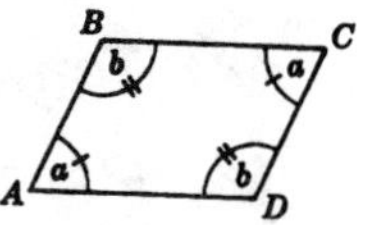

ALGEBRAIC PROOF:

Let a = number of degrees in $\angle A$ and $\angle C$,

b = number of degrees in $\angle B$ and $\angle D$.

$$2a + 2b = 360 \quad \text{(Pr. 3)}$$
$$a + b = 180$$

Since $\angle A$ and $\angle B$ are supplementary, $\overline{BC} \parallel \overline{AD}$.

Since $\angle A$ and $\angle D$ are supplementary, $\overline{AB} \parallel \overline{CD}$.

Chapter 5
Trapezoids and Parallelograms

IN THIS CHAPTER:

- ✔ *Trapezoids*
- ✔ *Parallelograms*
- ✔ *Special Parallelograms: Rectangle, Rhombus, Square*

Trapezoids

A *trapezoid* is a quadrilateral having two and only two parallel sides. The *bases* of the trapezoid are its parallel sides; the *legs* are its nonparallel sides. The *median* of the trapezoid is the segment joining the midpoints of its legs.

Thus, in trapezoid $ABCD$ in Fig. 5-1, the bases are $\overline{AD}$ and $\overline{BC}$, and the legs are $\overline{AB}$ and $\overline{CD}$. If M and N are midpoints, then $\overline{MN}$ is the median of the trapezoid.

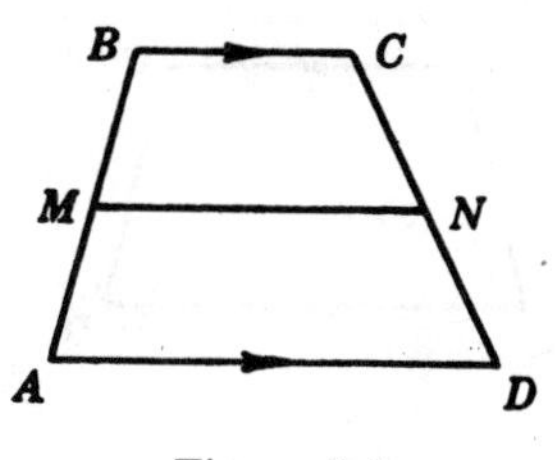

Figure 5-1

An *isosceles trapezoid* is a trapezoid whose legs are congruent. Thus, in isosceles trapezoid *ABCD* in Fig. 5-2, $\overline{AB} \cong \overline{CD}$. The *base angles* of a trapezoid are the angles at the ends of its longer base: $\angle A$ and $\angle D$ are the base angles of isosceles trapezoid *ABCD*.

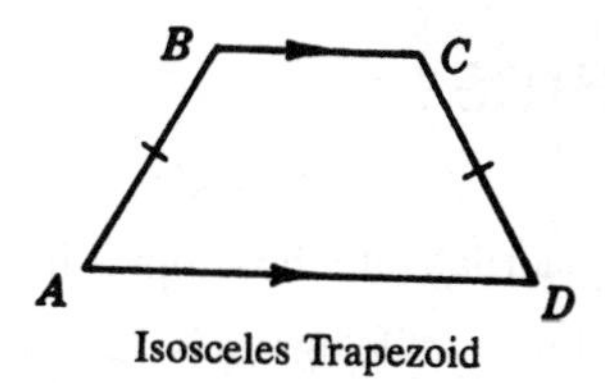

Figure 5-2

Trapezoid Principles

PRINCIPLE 1: *The base angles of an isosceles trapezoid are congruent.*

Thus, in trapezoid *ABCD* of Fig. 5-3, if $\overline{AB} \cong \overline{CD}$, then $\angle A \cong \angle D$.

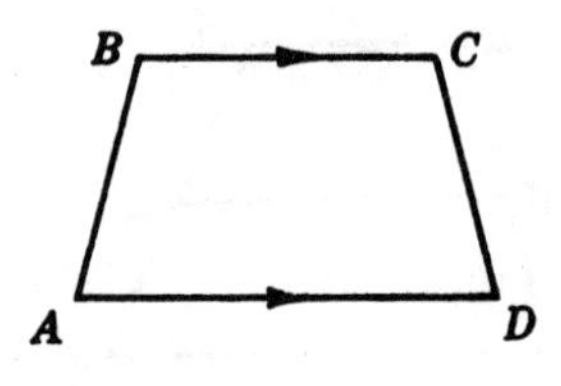

Figure 5-3

Principle 2: *If the base angles of a trapezoid are congruent, the trapezoid is isosceles.*

Thus, in Fig. 5-3, if $\angle A \cong \angle D$, then $\overline{AB} \cong \overline{CD}$.

Parallelograms

A parallelogram is a quadrilateral whose opposite sides are parallel. The symbol for parallelogram is ▱. Thus, in ▱*ABCD* in Fig. 5-4, $\overline{AB} \parallel \overline{CD}$ and $\overline{AD} \parallel \overline{BC}$. If the opposite sides of a quadrilateral are parallel, then it is a parallelogram. Thus, if $\overline{AB} \parallel \overline{CD}$ and $\overline{AD} \parallel \overline{BC}$, then *ABCD* is a ▱.

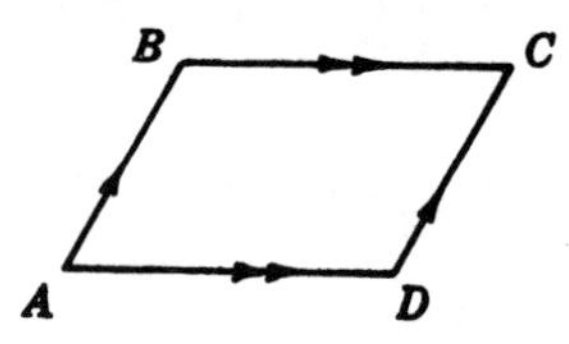

Figure 5-4

Principles Involving Properties of Parallelograms

PRINCIPLE 1: *The opposite sides of a parallelogram are parallel.*

PRINCIPLE 2: *A diagonal of a parallelogram divides it into two congruent triangles.*

PRINCIPLE 3: *The opposite sides of a parallelogram are congruent.*

PRINCIPLE 4: *The opposite angles of a parallelogram are congruent.*

PRINCIPLE 5: *The consecutive angles of a parallelogram are supplementary.*

PRINCIPLE 6: *The diagonals of a parallelogram bisect each other.*

Proving a Quadrilateral is a Parallelogram

PRINCIPLE 7: *A quadrilateral is a parallelogram if its opposite sides are parallel.*

PRINCIPLE 8: *A quadrilateral is a parallelogram if its opposite sides are congruent.*

PRINCIPLE 9: *A quadrilateral is a parallelogram if two sides are congruent and parallel.*

PRINCIPLE 10: *A quadrilateral is a parallelogram if its opposite angles are congruent.*

PRINCIPLE 11: *A quadrilateral is a parallelogram if its diagonals bisect each other.*

Special Parallelograms: Rectangle, Rhombus, Square

Definitions and Relationships among the Special Parallelograms

Rectangles, rhombuses, and squares belong to the set of parallelograms. Each of these may be defined as a parallelogram as follows:

1. A *rectangle* is an equiangular parallelogram.
2. A *rhombus* is an equilateral parallelogram.
3. A *square* is an equilateral and equiangular parallelogram.

Thus, a square is both a rectangle and a rhombus.

The relations among the special parallelograms can be pictured by using a circle to represent each set (Fig. 5-5).

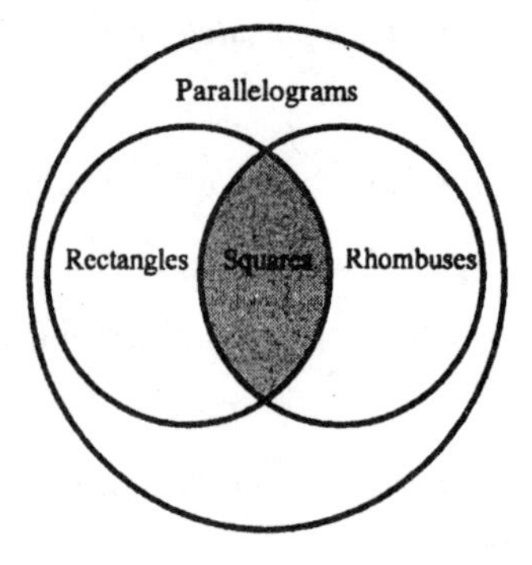

Figure 5-5

1. Since every rectangle and every rhombus must be a parallelogram, the circle for the set of rectangles and the circle for the set of rhombuses must be inside the circle for the set of parallelograms.

2. Since every square is both a rectangle and a rhombus, the overlapping shaded section must represent the set of squares.

Principles Involving Properties of the Special Parallelograms

PRINCIPLE 1: *A rectangle, rhombus, or square has all the properties of a parallelogram.*

PRINCIPLE 2: *Each angle of a rectangle is a right angle.*

PRINCIPLE 3: *The diagonals of a rectangle are congruent.*

PRINCIPLE 4: *All sides of a rhombus are congruent.*

PRINCIPLE 5: *The diagonals of a rhombus are perpendicular bisectors of each other.*

PRINCIPLE 6: *The diagonals of a rhombus bisect the vertex angles.*

PRINCIPLE 7: *The diagonals of a rhombus form four congruent triangles.*

PRINCIPLE 8: *A square has all the properties of both the rhombus and the rectangle.*

Diagonal Properties of Parallelograms, Rectangles, Rhombuses, and Squares

Each check in the following table indicates a diagonal property of the figure.

Diagonal Properties	Parallelogram	Rectangle	Rhombus	Square
Diagonals bisect each other.	✓	✓	✓	✓
Diagonals are congruent.		✓		✓
Diagonals are perpendicular.			✓	✓
Diagonals bisect vertex angles.			✓	✓
Diagonals form 2 pairs of congruent triangles.	✓	✓	✓	✓
Diagonals form 4 congruent triangles			✓	✓

Proving that a Parallelogram is a Rectangle, Rhombus, or a Square

Proving that a Parallelogram is a Rectangle

The basic or minimum definition of a rectangle is this: *A rectangle is a parallelogram having one right angle*. Since the consecutive angles of a parallelogram are supplementary, if one angle is a right angle, the remaining angles must be right angles.

The converse of this basic definition provides a useful method of proving that a parallelogram is a rectangle, as follows:

PRINCIPLE 9: *If a parallelogram has one right angle, then it is a rectangle.*

PRINCIPLE 10: *If a parallelogram has congruent diagonals, then it is a rectangle.*

Proving that a Parallelogram is a Rhombus

The basic or minimum definition of a rhombus is this: *A rhombus is a parallelogram having two congruent adjacent sides.*

The converse of this basic definition provides a useful method of proving that a parallelogram is a rhombus, as follows:

PRINCIPLE 11: *If a parallelogram has congruent adjacent sides, then it is a rhombus.*

Proving that a Parallelogram is a Square

PRINCIPLE 12: *If a parallelogram has a right angle and two congruent adjacent sides, then it is a square.*

This follows from the fact that a square is both a rectangle and a rhombus.

Solved Problem 5-1. Prove that a diagonal of a rhombus bisects each vertex angle through which it passes.

Solution

Given: Rhombus $ABCD$
$\overline{AC}$ is a diagonal.

To Prove: $\overline{AC}$ bisects $\angle A$ and $\angle C$.

Plan: Prove (*1*) $\angle 1$ and $\angle 2$ are congruent to $\angle 3$.
(*2*) $\angle 3$ and $\angle 4$ are congruent to $\angle 1$.

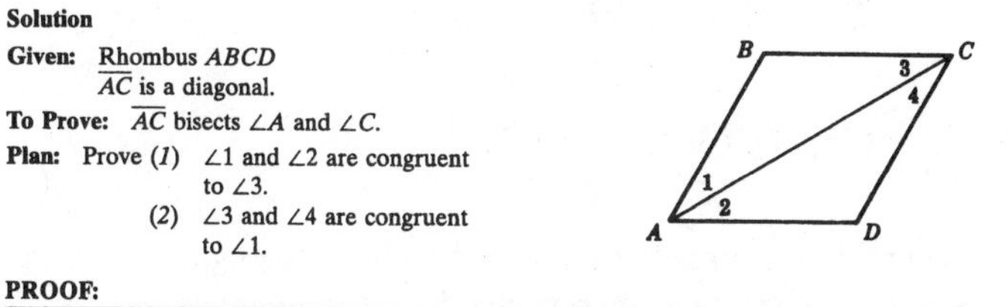

PROOF:

Statements	Reasons
1. $ABCD$ is a rhombus.	1. Given
2. $\overline{AB} \cong \overline{BC}$	2. A rhombus is equilateral.
3. $\angle 1 \cong \angle 3$	3. In a Δ, angles opposite congruent sides are congruent.
4. $\overline{BC} \parallel \overline{AD}$, $\overline{AB} \parallel \overline{CD}$	4. Opposite sides of a ▱ are $\parallel$.
5. $\angle 2 \cong \angle 3$, $\angle 1 \cong \angle 4$	5. Alternate interior ∡ of $\parallel$ lines are congruent.
6. $\angle 1 \cong \angle 2$, $\angle 3 \cong \angle 4$	6. Things congruent to the same thing are congruent to each other.
7. $\overline{AC}$ bisects $\angle A$ and $\angle C$.	7. To divide into two congruent parts is to bisect.

Chapter 6 CIRCLES

IN THIS CHAPTER:

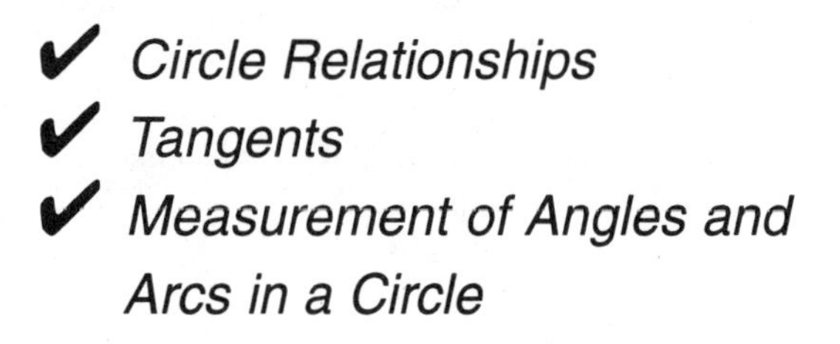

- ✔ *Circle Relationships*
- ✔ *Tangents*
- ✔ *Measurement of Angles and Arcs in a Circle*

Circle Relationships

The following terms are associated with a circle. Although some have been defined previously, they are repeated here for reference.

A *circle* is the set of all points in a plane that are the same distance from the *center.*

The *circumference* of a circle is the distance around the circle. It contains 360°.

A *radius* is a segment joining the center of a circle to a point on the circle (see Fig. 6-1).

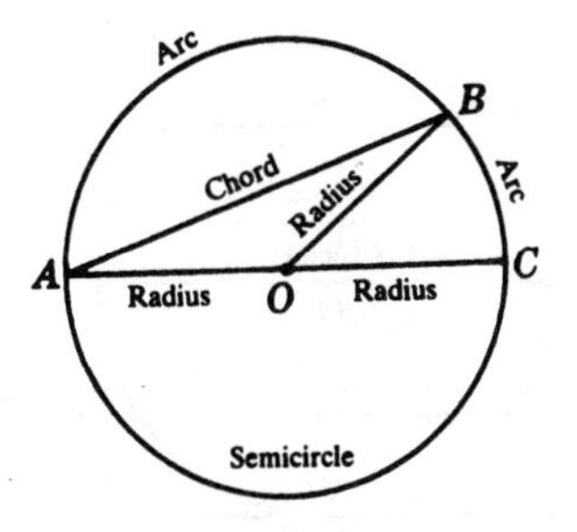

Figure 6-1

From the definition of a circle, it follows that the radii of a circle are congruent.

A *chord* is a segment joining any two points on a circle.

A *diameter* is a chord through the center of the circle; it is the longest chord and is twice the length of a radius.

A *central angle* is an angle formed by two radii.

An *arc* is a continuous part of a circle. A *semicircle* is an arc measuring one-half the circumference of a circle and thus contains 180°.

A *minor arc* is an arc that is less than a semicircle. A *major arc* is an arc that is greater than a semicircle. Thus, in Fig. 6-1, $\overset{\frown}{BC}$ is a minor arc and $\overset{\frown}{BAC}$ is a major arc.

To *intercept* an arc is to cut off the arc. Thus, in Fig. 6-1, $\angle BAC$ and $\angle BOC$ intercept $\overset{\frown}{BC}$.

A *chord* of a circle is a line segment joining two points of the circumference. Thus, in Fig. 6-2, $\overline{AB}$ is a chord.

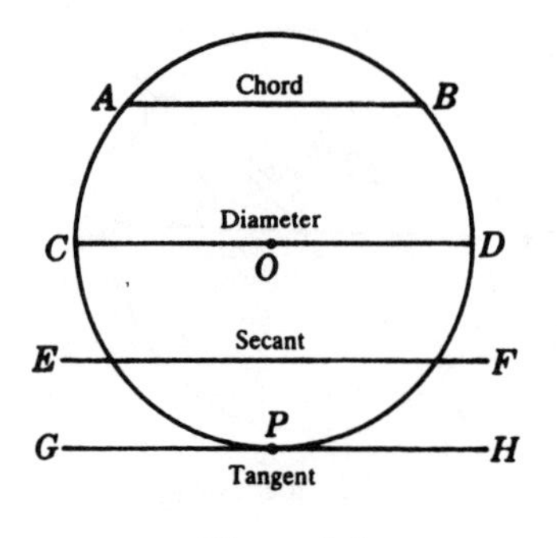

Figure 6-2

A *diameter* of a circle is a chord through the center. A *secant* of a circle is a line that intersects the circle at two points. A *tangent* of a circle is a line that touches the circle at one and only one point no matter how far produced. Thus, in Fig. 6-2, $\overline{CD}$ is a diameter of circle O, $\overleftrightarrow{EF}$ is a secant, and $\overleftrightarrow{GH}$ is a tangent to the circle at P. P is the point of contact or the point of tangency.

An *inscribed polygon* is a polygon all of whose sides are chords of a circle. A *circumscribed* circle is a circle passing through each vertex of a polygon. Thus, $\triangle ABD$, $\triangle BCD$ and quadrilateral $ABCD$ are inscribed polygons of circle O in Fig. 6-3. Circle O is a circumscribed circle of quadrilateral $ABCD$.

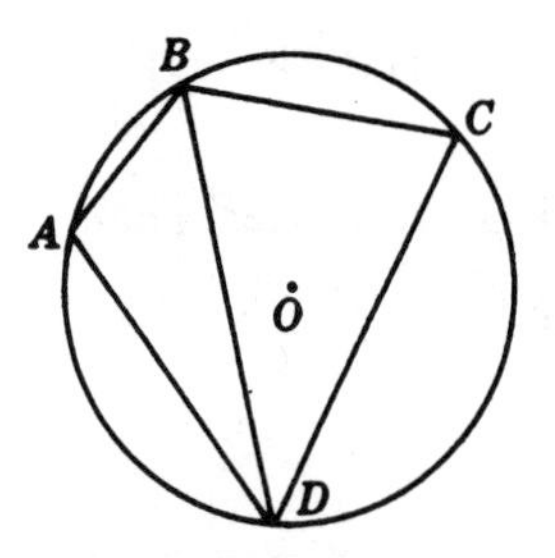

Inscribed Polygons
Circumscribed Circle

Figure 6-3

A *circumscribed polygon* is a polygon all of whose sides are tangents to a circle. An *inscribed circle* is a circle to which all sides of a polygon are tangents. Thus, $\triangle ABC$ is a circumscribed polygon of circle O in Fig. 6-4. Circle O is an inscribed circle of $\triangle ABC$.

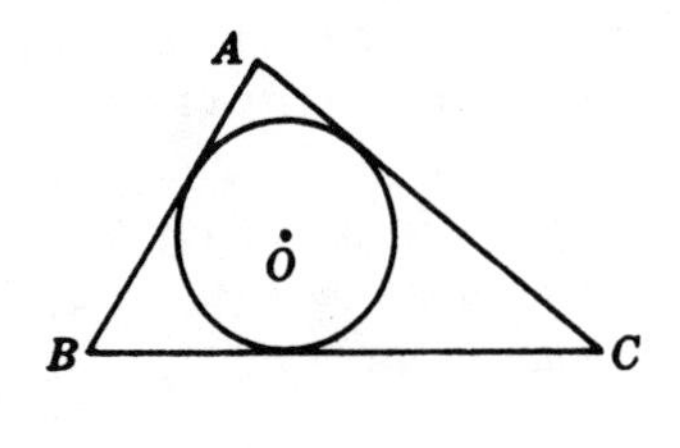

Circumscribed Polygon
Inscribed Circle
Figure 6-4

Concentric circles are circles that have the same center. Thus, the two circles shown in Fig. 6-5 are concentric circles, $\overline{AB}$ is a tangent of the inner circle and a chord of the outer one. $\overline{CD}$ is a secant of the inner circle and a chord of the outer one.

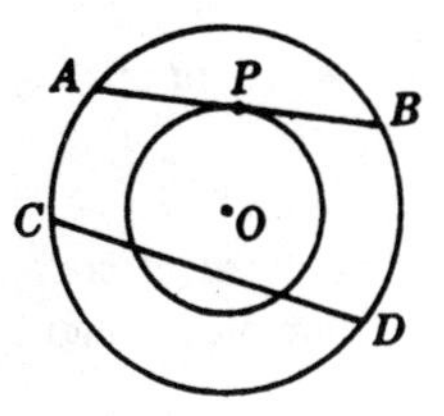

Concentric Circles
Figure 6-5

Two circles are *equal* if their radii are equal in length; two circles are *congruent* if their radii are congruent.

Two arcs are congruent if they have equal degree measure and length. We use the notation $m\,\overset{\frown}{AC}$ to denote "measure of arc AC."

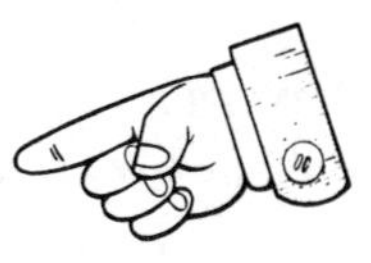

Circle Principles

Principle 1: *A diameter divides a circle into two equal parts.*

Principle 2: *If a chord divides a circle into two equal parts, then it is a diameter.* (This is the converse of Principle 1.)

Principle 3: *A point is outside, on, or inside a circle according to whether its distance from the center is greater than, equal to, or smaller than the radius.*

Principle 4: *Radii of the same or congruent circles are congruent.*

Principle 5: *Diameters of the same or congruent circles are congruent.*

Principle 6: *In the same or congruent circles, congruent central angles have congruent arcs.*

Principle 7: *In the same or congruent circles, congruent arcs have congruent central angles.*

Principle 8: *In the same or congruent circles, congruent chords have congruent arcs.*

Principle 9: *In the same or congruent circles, congruent arcs have congruent chords.* (Principles 8 and 9 are converses of each other.)

Principle 10: *A diameter perpendicular to a chord bisects the chord and its arcs.*

Principle 11: *A perpendicular bisector of a chord passes through the center of the circle.*

Principle 12: *In the same or congruent circles, congruent chords are equally distant from the center.*

Principle 13: *In the same or congruent circles, chords that are equally distant from the center are congruent.* (Principles 12 and 13 are converses of each other.)

Tangents

The *length of a tangent* from a point to a circle is the length of the segment of the tangent from the given point to the point of tangency. Thus, *PA* is the length of the tangent from *P* to circle *O* in Fig. 6-6.

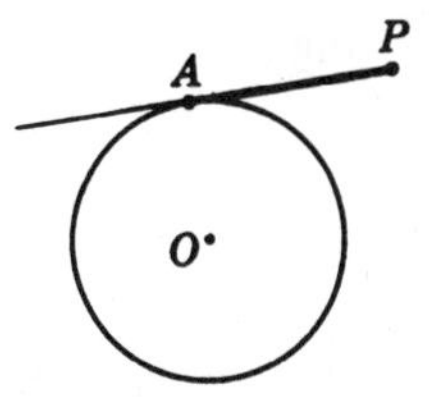

Figure 6-6

Tangent Principles

Principle 1: *A tangent is perpendicular to the radius drawn to the point of contact.*

Principle 2: *A line is tangent to a circle if it is perpendicular to a radius at its outer end.*

Principle 3: *A line passes through the center of a circle if it is perpendicular to a tangent at its point of contact.*

Principle 4: *Tangents to a circle from an outside point are congruent.*

Principle 5: *The segment from the center of a circle to an outside point bisects the angle between the tangents from the point to the circle.*

Two Circles in Varying Relative Positions

The *line of centers of two circles* is the line joining their centers.

Thus $\overline{OO'}$ is the line of centers of circles O and O' in Fig. 6-7.

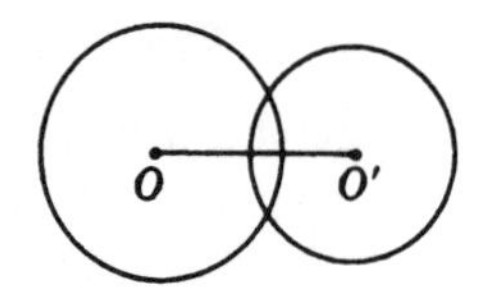

Figure 6-7

Circles Tangent Externally

Circles O and O' in Fig. 6-8 are tangent externally at P. $\overleftrightarrow{AB}$ is the common internal tangent of both circles. The line of centers $\overline{OO'}$ passes through P, is perpendicular to $\overleftrightarrow{AB}$, and is equal in length to the sum of the radii, $R + r$. Also, $\overleftrightarrow{AB}$ bisects each of the common external tangents, $\overline{CD}$ and $\overline{EF}$.

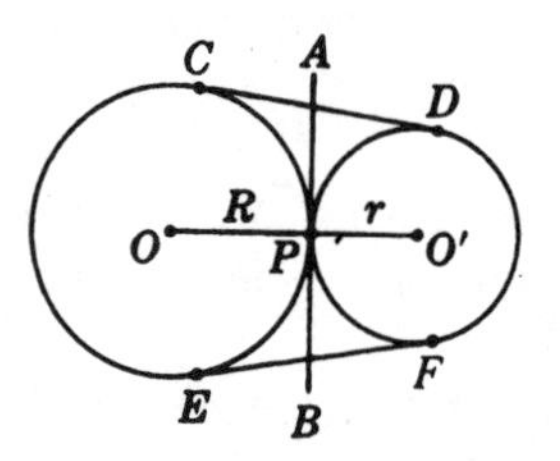

Figure 6-8

Circles Tangent Internally

Circles O and O' in Fig. 6-9 are tangent internally at P. $\overleftrightarrow{AB}$ is the common external tangent of both circles. The line of centers $\overline{OO'}$ if extended, passes through P, is perpendicular to $\overleftrightarrow{AB}$, and is equal in length to the difference of the radii, $R - r$.

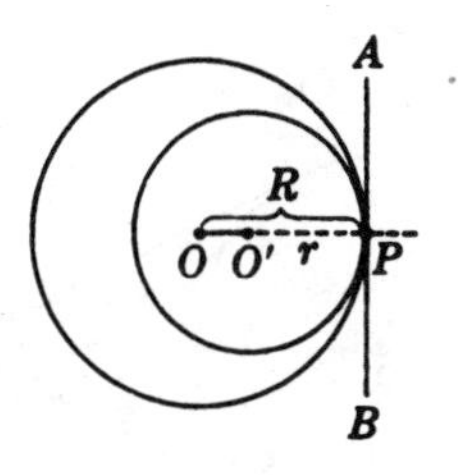

Figure 6-9

Overlapping Circles

Circles O and O' in Fig. 6-10 overlap. Their common chord is $\overline{AB}$. If the circles are unequal, their (equal) common external tangents $\overleftrightarrow{CD}$ and $\overleftrightarrow{EF}$ meet at P. The line of centers $\overline{OO'}$ is the perpendicular bisector of $\overline{AB}$ and, if extended, passes through P.

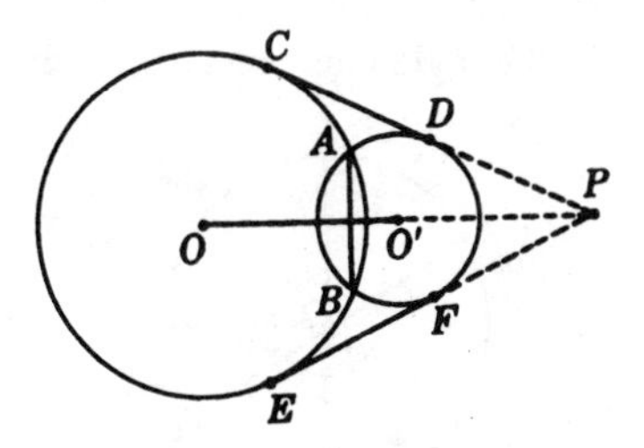

Figure 6-10

Circles Outside Each Other

Circles O and O' in Fig. 6-11 are entirely outside each other. The common internal tangents, $\overline{AB}$ and $\overline{CD}$, meet at P. If the circles are unequal, their common external tangents, $\overline{EF}$ and $\overline{GH}$, if extended, meet at P'. The line of centers $\overline{OO'}$ passes through P and P'. Also, $\overline{AB} = \overline{CD}$ and $\overline{EF} = \overline{GH}$.

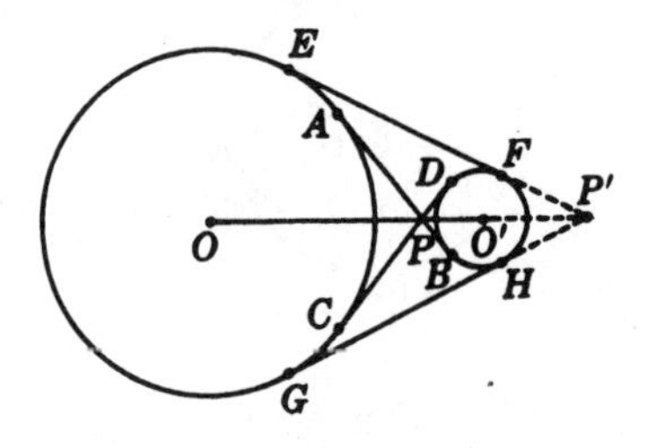

Figure 6-11

Measurements of Angles and Arcs in a Circle

A *central angle* has the same number of degrees as the arc it intercepts. Thus, as shown in Fig. 6-12, a central angle which is a right angle intercepts a 90° arc; a 40° central angle intercepts a 40° arc; and a central angle which is a straight angle intercepts a semicircle of 180°.

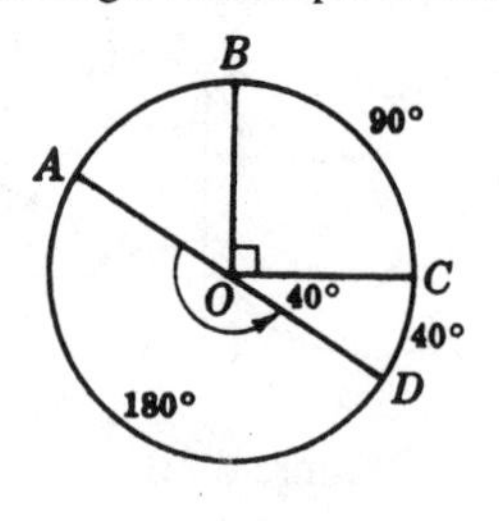

Figure 6-12

An *inscribed angle* is an angle whose vertex is on the circle and whose sides are chords. An *angle inscribed in an arc* has its vertex on the arc and its sides passing through the ends of the arc. Thus, $\angle A$ intercepts $\widehat{BC}$ and is inscribed in $\widehat{BAC}$.

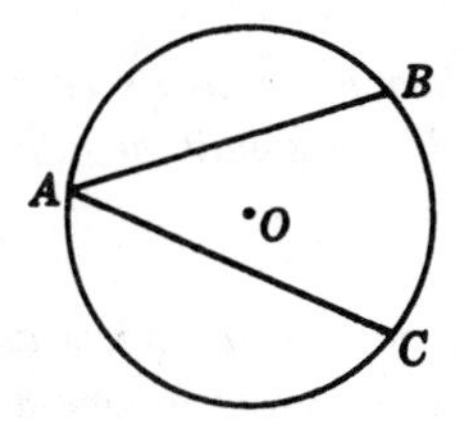

Figure 6-13

Angle-Measurement Principles

Principle 1: *A central angle is measured by its intercepted arc.*

Principle 2: *An inscribed angle is measured by one-half its intercepted arc.*

Principle 3: *In the same or congruent circles, congruent inscribed angles have congruent intercepted arcs.*

Principle 4: *In the same or congruent circles, inscribed angles having congruent intercepted arcs are congruent.*

Principle 5: *Angles inscribed in the same or congruent arcs are congruent.*

Principle 6: *An angle inscribed in a semicircle is a right angle.*

Principle 7: *Opposite angles of an inscribed quadrilateral are supplementary.*

Principle 8: *Parallel lines intercept congruent arcs on a circle.*

Principle 9: *An angle formed by a tangent and a chord is measured by one-half its intercepted arc.*

Principle 10: *An angle formed by two intersecting chords is measured by one-half the sum of the intercepted arcs.*

Principle 11: *An angle formed by two secants intersecting outside a circle is measured by one-half the difference of the intercepted arcs.*

Principle 12: *An angle formed by a tangent and a secant intersecting outside a circle is measured by one-half the difference of the intercepted arcs.*

Principle 13: *An angle formed by two tangents intersecting outside a circle is measured by one-half the difference of the intercepted arcs.*

Table of Angle-Measurement Principles

Position of Vertex	Kind of Angle	Diagram	Measurement Formula	Method of Measurement
Center of circle	Central angle (apply Principle 1)		$\angle O \doteq AB$ $m\angle O = a^\circ$	By intercepted arc
On the circle	Inscribed angle (apply Principle 2)		$\angle A \doteq \frac{1}{2}\widehat{BC}$ $m\angle A = \frac{1}{2}a^\circ$	By one-half intercepted arc
	Angle formed by a tangent and a chord (apply Principle 9)			

Table of Angle-Measurement Principles (*Cont.*)

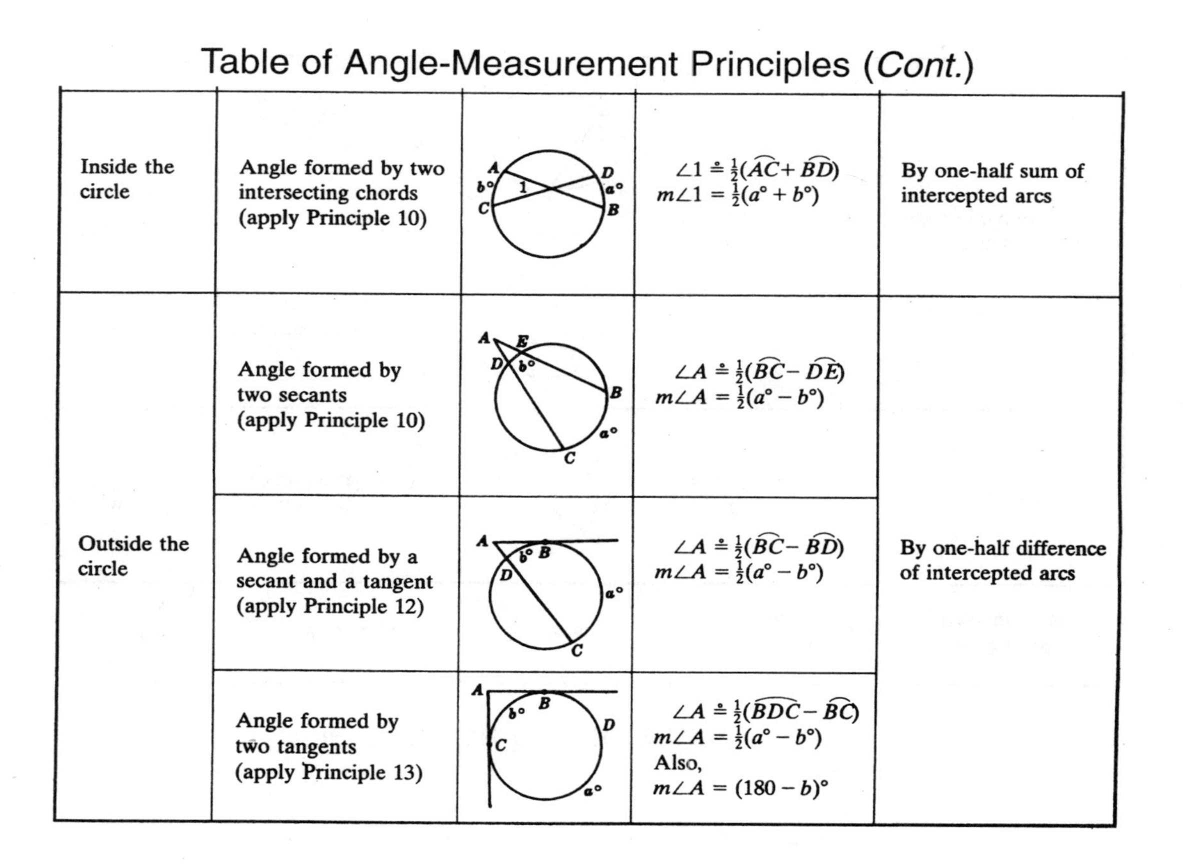

Inside the circle	Angle formed by two intersecting chords (apply Principle 10)		$\angle 1 \doteq \frac{1}{2}(\widehat{AC} + \widehat{BD})$ $m\angle 1 = \frac{1}{2}(a^\circ + b^\circ)$	By one-half sum of intercepted arcs
Outside the circle	Angle formed by two secants (apply Principle 10)		$\angle A \doteq \frac{1}{2}(\widehat{BC} - \widehat{DE})$ $m\angle A = \frac{1}{2}(a^\circ - b^\circ)$	By one-half difference of intercepted arcs
	Angle formed by a secant and a tangent (apply Principle 12)		$\angle A \doteq \frac{1}{2}(\widehat{BC} - \widehat{BD})$ $m\angle A = \frac{1}{2}(a^\circ - b^\circ)$	
	Angle formed by two tangents (apply Principle 13)		$\angle A \doteq \frac{1}{2}(\widehat{BDC} - \widehat{BC})$ $m\angle A = \frac{1}{2}(a^\circ - b^\circ)$ Also, $m\angle A = (180 - b)^\circ$	

Solved Problem 6-1. Prove that parallel chords drawn at the ends of a diameter are equal in length.

Solutions

Given: Circle O
$\overline{AB}$ is a diameter.
$\overline{AC} \parallel \overline{BD}$

To Prove: $AC = BD$

Plan: Prove $\widehat{AC} \cong \widehat{BD}$

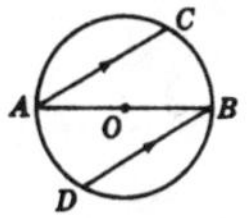

PROOF:

Statements	Reasons
1. $\overline{AB}$ is a diameter.	1. Given
2. $\widehat{ACB} \cong \widehat{ADB}$	2. A diameter cuts a circle into two equal semicircles.
3. $\overline{AC} \parallel \overline{BD}$	3. Given
4. $\overline{AD} \cong \overline{BC}$	4. Parallel lines intercept ≅ arcs on a circle.
5. $\overline{AC} \cong \overline{BD}$	5. If equals are subtracted from equals, the differences are equal. Definition of ≅ arcs.
6. $AC = BD$	6. In a circle, equal arcs have chords which are equal in length.

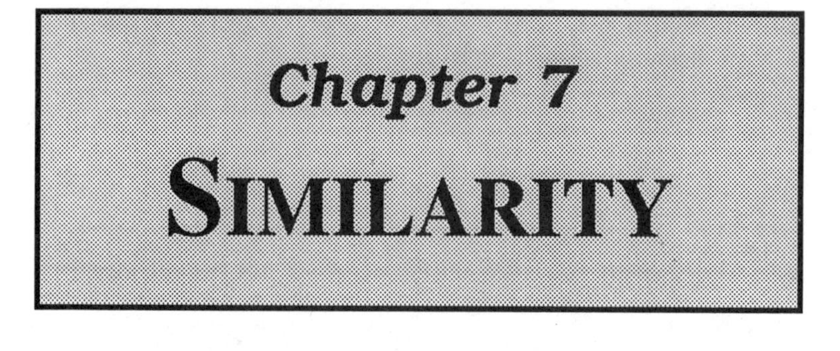

IN THIS CHAPTER:

- ✔ *Ratios*
- ✔ *Proportions*
- ✔ *Proportional Segments*
- ✔ *Similar Triangles*
- ✔ *Mean Proportionals in a Right Triangle*
- ✔ *Pythagorean Theorem*
- ✔ *Special Right Triangles*

Ratios

Ratios are used to compare quantities by division: the ratio of two quantities is the first divided by the second. A ratio is an abstract number, that is, a number without a unit of measure. Thus, the ratio of 10 ft to 5 ft is 10 ft ÷ 5 ft, which equals 2.

A ratio can be expressed in the following ways:

1. using a colon, as in 3:4
2. using "to" as in 3 to 4
3. as a common fraction, as in $\frac{3}{4}$
4. as a decimal, 0.75
5. as a percent, 75%.

The quantities involved in a ratio must have the same unit. A ratio should be simplified by reducing to lowest terms and eliminating fractions. Thus, to find the ratio of 1 ft to 4 in, we first change the foot to 12 inches, and then take the ratio of 12 inches to 4 inches; the result is a ratio of 3 to 1, or 3. Also, the ratio of $2\frac{1}{2}:\frac{1}{2}$ would be restated as 5:1 or 5.

The ratio of three or more quantities may be expressed as a *continued ratio*. Thus the ratio of $2 to $3 to $5 is the continued ratio 2:3:5. This enlarged ratio is a combination of three separate ratios; these are 2:3, 3:5, and 2:5.

Proportions

A *proportion* is an equality of two ratios. Thus, 2:5 = 4:10 (or $\frac{2}{5}=\frac{4}{10}$) is a proportion.

The fourth term of a proportion is the *fourth proportional* to the other three taken in order. Thus, in $2:3 = 4:x$, x is the fourth proportional to 2, 3, and 4.

The *means* of a proportion are its middle terms, that is, its second and third terms. The *extremes* of a proportion are its outside terms, that is, its first and fourth terms. Thus, in $a:b = c:d$, the means are b and c, and the extremes are a and d.

Proportion Principles

PRINCIPLE 1: *In any proportion, the product of the means equals the product of the extremes.*

PRINCIPLE 2: *If the product of two numbers equals the product of two other numbers, either pair may be made the means of a proportion and the other pair may be made the extremes.*

Methods of Changing a Proportion Into an Equivalent Proportion

Principle 3: (Inversion method) *A proportion may be changed into an equivalent proportion by inverting each ratio.*

Principle 4: (Alternation method) *A proportion may be changed into an equivalent proportion by interchanging the means or by interchanging the extremes.*

Principle 5: (Addition method) *A proportion may be changed into an equivalent proportion by adding terms in each ratio to obtain new first and third terms.*

Principle 6: (Subtraction method) *A proportion may be changed into an equivalent proportion by subtracting terms in each ratio to obtain new first and third terms.*

Other Proportion Principles

Principle 7: *If any three terms of one proportion equal the corresponding three terms of another proportion, the remaining terms are equal.*

Principle 8: *In a series of equal ratios, the sum of any of the numerators is to the sum of the corresponding denominators as any numerator is to its denominator.*

Proportional Segments

If two segments are divided proportionately, (1) the corresponding new segments are in proportion, and (2) the two original segments and either pair of corresponding new segments are in proportion. Thus, if $\overline{AB}$ and $\overline{AC}$ in Fig. 7-1 are divided proportionately by $\overline{DE}$, we may write a proportion such as $\frac{a}{b}=\frac{c}{d}$ using the four segments; or we may write a proportion such as $\frac{a}{AB}=\frac{c}{AC}$ using the two original segments and two of their new segments.

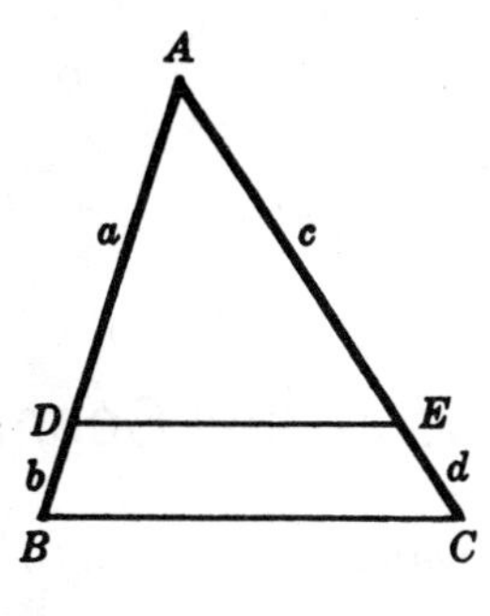

Figure 7-1

Obtaining the Eight Arrangements of any Proportion

A proportion such as $\frac{a}{b}=\frac{c}{d}$ can be arranged in eight ways. To obtain the eight variations, we let each term of the proportion represent one of the new segments of Fig. 7-1. Two of the possible proportions are then obtained from each direction, as follows.

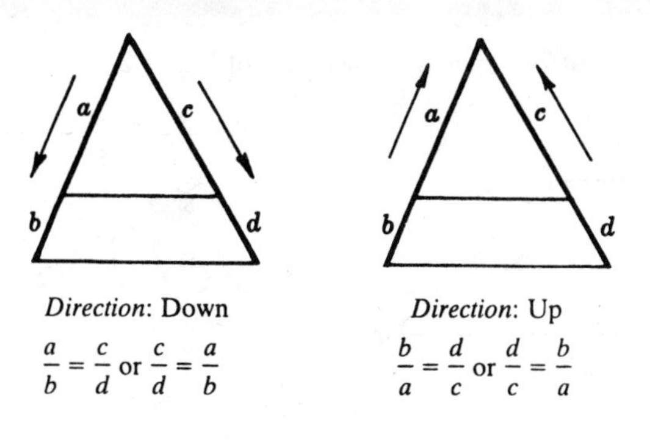

Direction: Down

$\frac{a}{b}=\frac{c}{d}$ or $\frac{c}{d}=\frac{a}{b}$

Direction: Up

$\frac{b}{a}=\frac{d}{c}$ or $\frac{d}{c}=\frac{b}{a}$

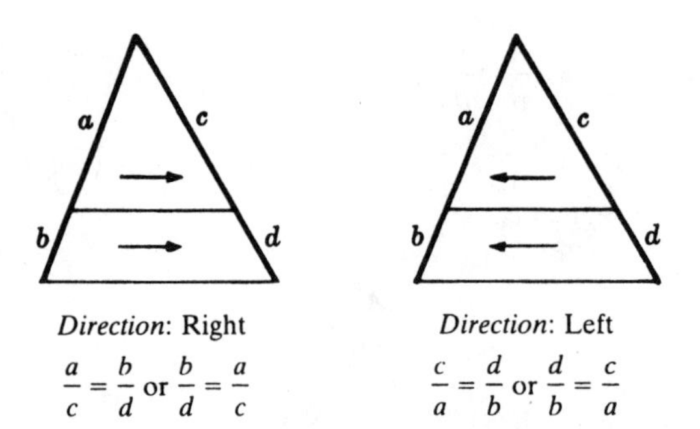

Direction: Right $\frac{a}{c} = \frac{b}{d}$ or $\frac{b}{d} = \frac{a}{c}$

Direction: Left $\frac{c}{a} = \frac{d}{b}$ or $\frac{d}{b} = \frac{c}{a}$

Principles of Proportional Segments

PRINCIPLE 1: *If a line is parallel to one side of a triangle, then it divides the other two sides proportionately.*

PRINCIPLE 2: *If a line divides two sides of a triangle proportionately, it is parallel to the third side.* (Principles 1 and 2 are converses.)

PRINCIPLE 3: *Three or more parallel lines divide any two transversals proportionately.*

PRINCIPLE 4: *A bisector of an angle of a triangle divides the opposite side into segments which are proportional to the adjacent sides.*

Solved Problem 7-1. Find x in each part of Fig. 7-2.

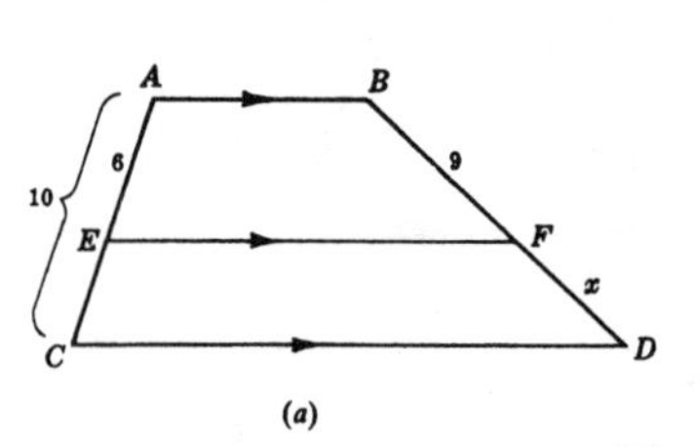

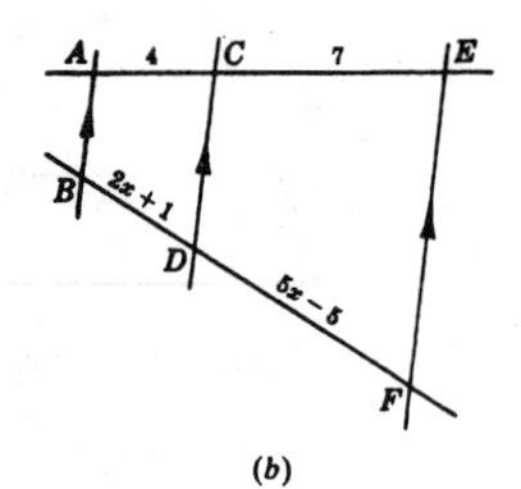

Figure 7-2

Solutions

(a) We have $EC = 4$ and $\overline{AB} \parallel \overline{EF} \parallel \overline{CD}$; hence $\frac{x}{9} = \frac{4}{6}$ and $x = 6$.

(b) $\overleftrightarrow{AB} \parallel \overleftrightarrow{CD} \parallel \overleftrightarrow{EF}$; hence $\frac{5x-5}{2x+1} = \frac{7}{4}$, from which $20x - 20 = 14x + 7$. Then $6x = 27$ and $x = 4\frac{1}{2}$.

Similar Triangles

Similar polygons are polygons whose corresponding angles are congruent and whose corresponding sides are in proportion. Similar polygons have the same shape although not necessarily the same size.

The symbol for "similar" is $\sim$. As in the case of congruent triangles, *corresponding sides of similar triangles are opposite congruent angles.*

In Fig. 7-3, $\triangle ABC \sim \triangle A'B'C'$ because

$$m\angle A = m\angle A' = 37^\circ \quad m\angle B = m\angle B' = 53^\circ \quad m\angle C = m\angle C' = 90^\circ$$

and $\frac{a}{a'} = \frac{b}{b'} = \frac{c}{c'}$ or $\frac{6}{3} = \frac{8}{4} = \frac{10}{5}$

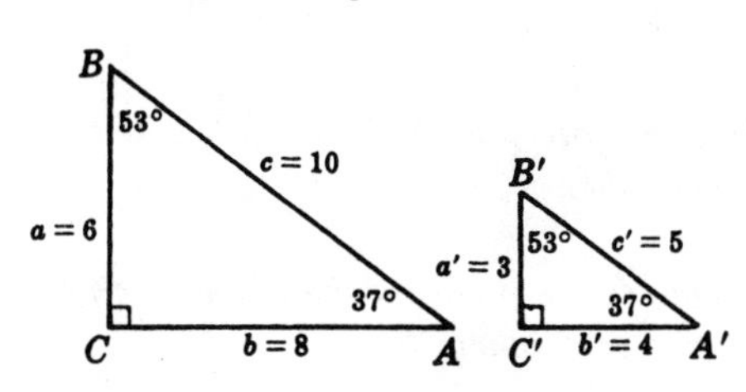

Figure 7-3

Principles of Similar Triangles

PRINCIPLE 1: *Corresponding angles of similar triangles are congruent.* (By the definition)

PRINCIPLE 2: *Corresponding sides of similar triangles are in proportion.* (By the definition)

PRINCIPLE 3: *Two triangles are similar if two angles of one triangle are congruent respectively to two angles of the other.*

PRINCIPLE 4: *Two triangles are similar if an angle of one triangle is congruent to an angle of the other and the sides including these angles are in proportion.*

PRINCIPLE 5: *Two triangles are similar if their corresponding sides are in proportion.*

PRINCIPLE 6: *Two right triangles are similar if an acute angle of one is congruent to an acute angle of the other.* (Corollary of Principle 3)

PRINCIPLE 7: *A line parallel to a side of a triangle cuts off a triangle similar to the given triangle.*

PRINCIPLE 8: *Triangles similar to the same triangle are similar to each other.*

PRINCIPLE 9: *The altitude to the hypotenuse of a right triangle divides it into two triangles which are similar to the given triangle and to each other.*

PRINCIPLE 10: *Triangles are similar if their sides are respectively parallel to each other.*

PRINCIPLE 11: *Triangles are similar if their sides are respectively perpendicular to each other.*

Solved Problem 7-2. Prove that two isosceles triangles are similar if a base angle of one is congruent to a base angle of the other.

Given: Isosceles $\triangle ABC$ ($AB = AC$)
Isosceles $\triangle A'B'C'$ ($A'B' = A'C'$)
$\angle B \cong \angle B'$
To Prove: $\triangle ABC \sim \triangle A'B'C'$
Plan: Prove $\angle C \cong \angle C'$ and use Principle 3.

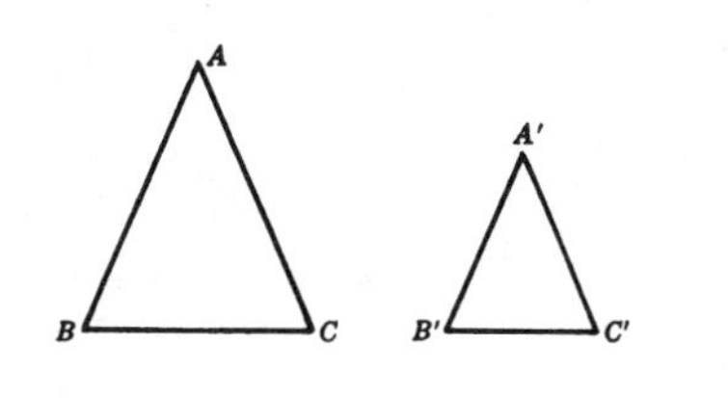

PROOF:

Statements	Reasons
1. $\angle B \cong \angle B'$	1. Given
2. $\angle B \cong \angle C$, $\angle B' \cong \angle C'$	2. Base angles of an isosceles triangle are congruent.
3. $\angle C \cong \angle C'$	3. Things $\cong$ to $\cong$ things are $\cong$ to each other.
4. $\triangle ABC \sim \triangle A'B'C'$	4. Two triangles are similar if two angles of one triangle are congruent to two angles of the other.

Mean Proportionals in a Right Triangle

PRINCIPLE 1: *The length of the altitude to the hypotenuse of a right triangle is the mean proportional between the lengths of the segments of the hypotenuse.*

PRINCIPLE 2: *In a right triangle, the length of either leg is the mean proportional between the length of the hypotenuse and the length of the projection of that leg on the hypotenuse.*

Pythagorean Theorem

In a right triangle, the square of the length of the hypotenuse equals the sum of the squares of the lengths of the legs. Thus, in Fig. 7-4, $c^2 = a^2 + b^2$.

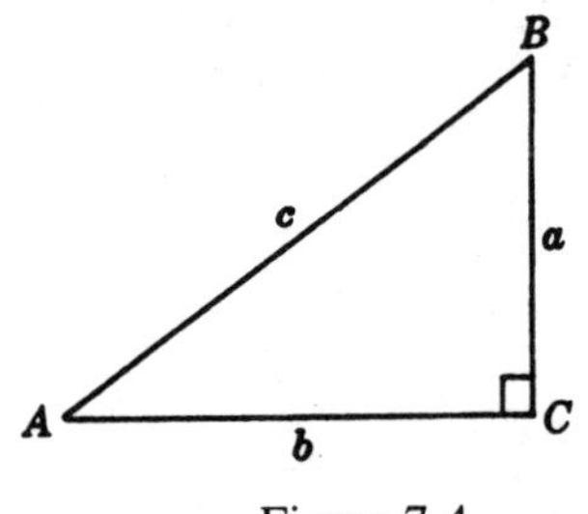

Figure 7-4

Tests for Right, Acute, and Obtuse Triangles

If $c^2 = a^2 + b^2$ applies to the three sides of a triangle, then the triangle is a right triangle; but if $c^2 \neq a^2 + b^2$, then the triangle is not a right triangle.

In $\triangle ABC$, if $c^2 < a^2 + b^2$ where c is the longest side of the triangle, then the triangle is an acute triangle. Thus, in Fig. 7-5, $9^2 < 6^2 + 8^2$ (that is, $81 < 100$); hence $\triangle ABC$ is an acute triangle.

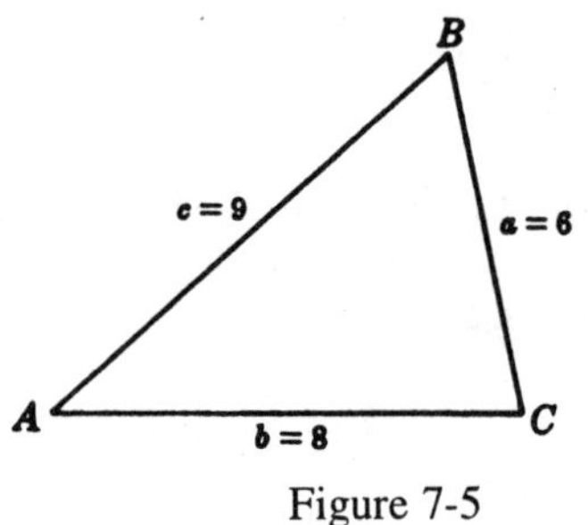

Figure 7-5

In $\triangle ABC$, if $c^2 > a^2 + b^2$ where c is the longest side of the triangle, then the triangle is an obtuse triangle. Thus, in Fig. 7-6, $11^2 > 6^2 + 8^2$ (that is, $121 > 100$); hence $\triangle ABC$ is an obtuse triangle.

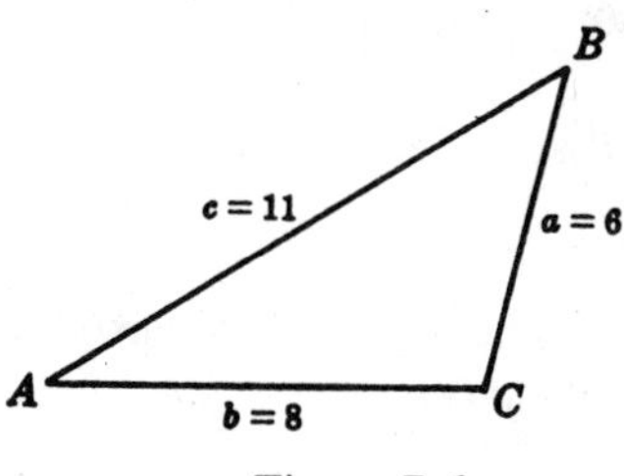

Figure 7-6

Special Right Triangles

The 30°-60°-90° Triangle

A 30°-60°-90° triangle is one-half an equilateral triangle. Thus, in right $\triangle ABC$ (Fig. 7-7), $a = \frac{1}{2}c$. Consider that $c = 2$; then $a = 1$, and the Pythagorean Theorem gives

$$b^2 = c^2 - a^2 = 2^2 - 1^2 = 3 \text{ or } b = \sqrt{3}$$

The ratio of the sides is then $a: b: c = 1: \sqrt{3} : 2$.

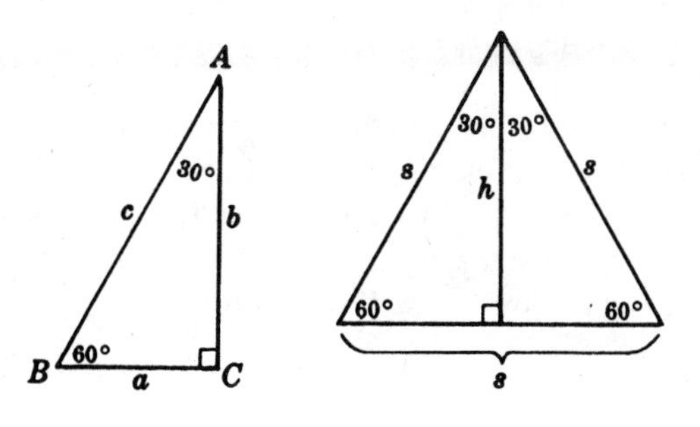

Figure 7-7

Principles of the 30°-60°-90° Triangle

PRINCIPLE 1: *The length of the leg opposite the* 30° *angle equals one-half the length of the hypotenuse.*

PRINCIPLE 2: *The length of the leg opposite the* 60° *angle equals one-half the length of the hypotenuse times the square root of* 3.

PRINCIPLE 3: *The length of the leg opposite the* 60° *angle equals the length of the leg opposite the* 30° *angle times the square root of* 3.

PRINCIPLE 4: *The length of the altitude of an equilateral triangle equals one-half the length of a side times the square root of* 3. (Principle 4 is a corollary of Principle 2.)

Principles of the 45°-45°-90° Triangle

PRINCIPLE 5: *The length of a leg opposite a* 45° *angle equals one-half the length of the hypotenuse times the square root of* 2.

PRINCIPLE 6: *The length of the hypotenuse equals the length of a side times the square root of* 2.

PRINCIPLE 7: *In a square, the length of a diagonal equals the length of a side times the square root of* 2.

Solved Problem 7-3. (a) If the length of the hypotenuse of a 30°-60°-90° triangle is 12, find the lengths of its legs [Fig. 7-8(a)]. (b) Each leg of an isosceles trapezoid has length 18. If the base angles are 60° and the upper base is 10, find the lengths of the altitude and the lower base [Fig. 7-8(b)].

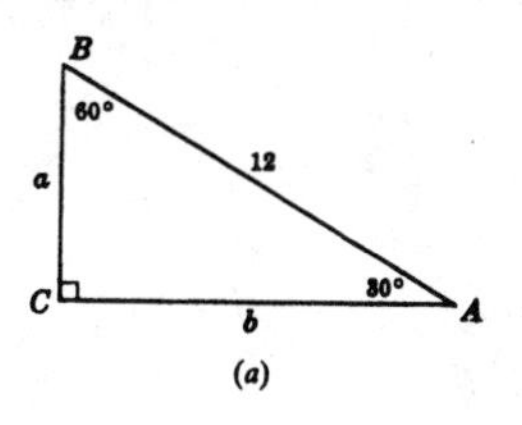

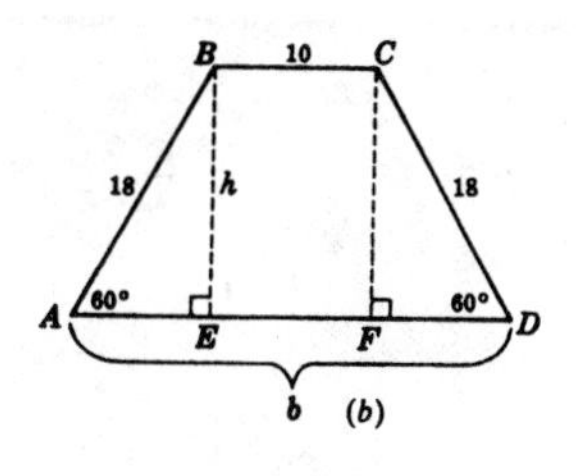

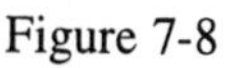
Figure 7-8

Solutions

(a) By Principle 1, $a = \frac{1}{2}(12) = 6$. By Principle 2, $b = \frac{1}{2}(12)\sqrt{3} = 6\sqrt{3}$.

(b) By Principle 2, $h = \frac{1}{2}(18)\sqrt{3} = 9\sqrt{3}$. By Principle 1, $\overline{AE} = \overline{FD} = \frac{1}{2}(18) = 9$; hence $b = 9 + 10 + 9 = 28$.

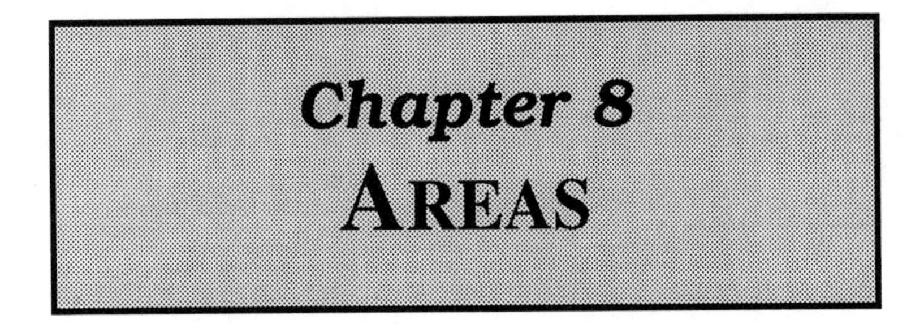

IN THIS CHAPTER:

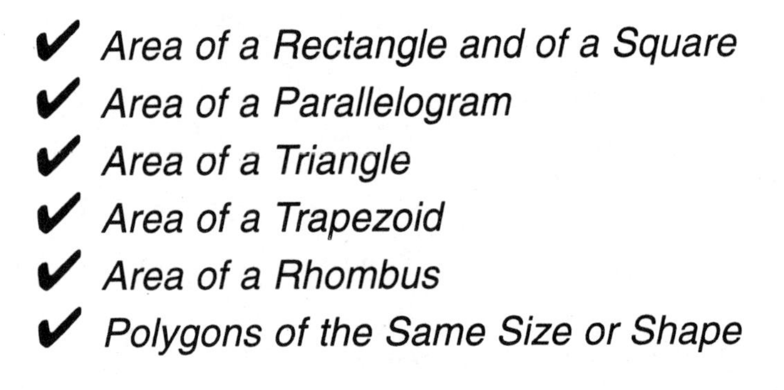

Area of a Rectangle and of a Square

A *square* unit is the surface enclosed by a square whose side is 1 unit (Fig. 8-1). The *area of a closed plane figure*, such as a polygon, is the number of square units contained in its surface. Since a rectangle 5 units long and 4 units wide can be divided into 20 unit squares, its area is 20 square units (Fig. 8-2).

The *area of a rectangle equals the product of the length of its base and the length of its altitude (height)* (Fig. 8-3).

The *area of a square equals the square of the length of a side* (Fig. 8-4).

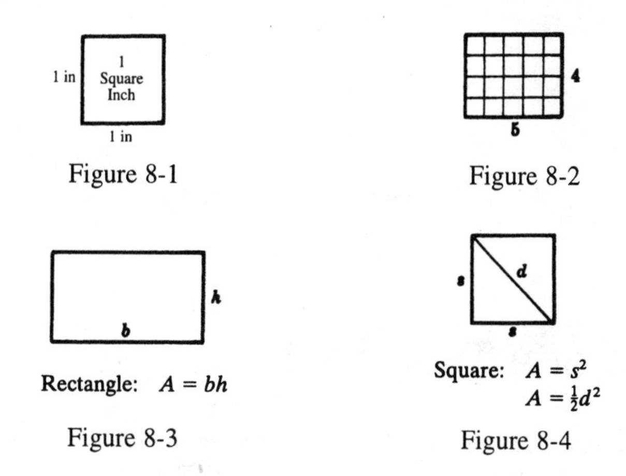

Figure 8-1

Figure 8-2

Figure 8-3

Figure 8-4

Area of a Parallelogram

The *area of a parallelogram equals the product of the length of a side and the length of the altitude to that side*. Thus, in $\square ABCD$ (Fig. 8-5), if $b = 10$ and $h = 2.7$, then $A = 10(2.7) = 27$.

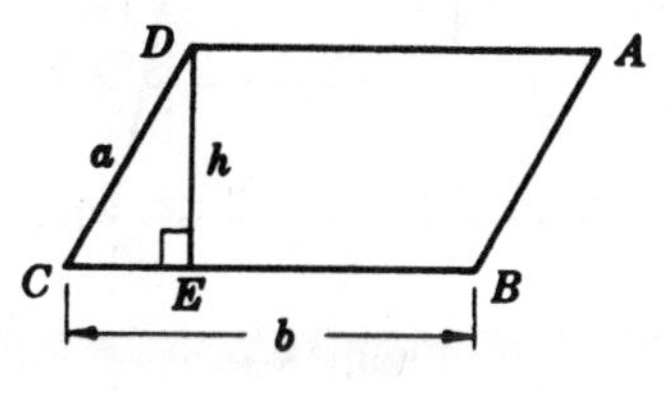

Figure 8-5

Area of a Triangle

The *area of a triangle equals one-half the product of the length of a side and the length of the altitude to that side*, as shown in Fig. 8-6.

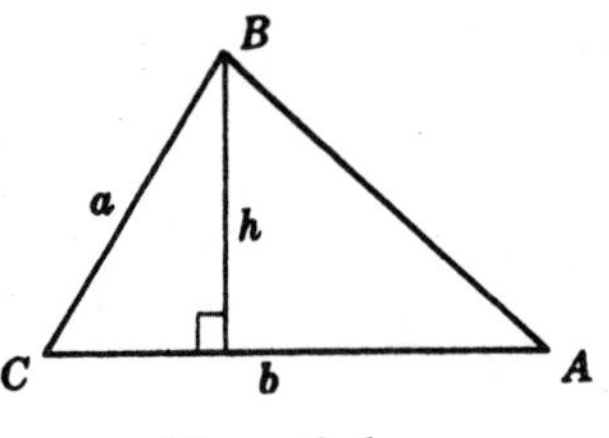

Figure 8-6

Area of a Trapezoid

The area of a trapezoid equals one-half the product of the length of its altitude and the sum of the lengths of its bases. Thus, if $h = 20$, $b = 27$, and $b' = 23$ in Fig. 8-7, then $A = \frac{1}{2}(20)(27+23) = 500$.

The area of a trapezoid equals the product of the lengths of its altitude and median.

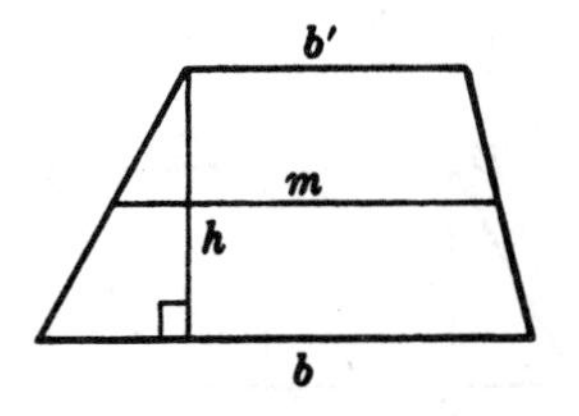

Figure 8-7

Area of a Rhombus

The area of a rhombus equals one-half the product of the lengths of its diagonals. Since each diagonal is the perpendicular bisector of the other, the area of triangle I in Fig. 8-8 is $\frac{1}{2}\left(\frac{1}{2}d\right)\left(\frac{1}{2}d'\right) = \frac{1}{8}dd'$. Thus, the rhombus, which consists of four triangles congruent to $\triangle$ I, has an area of $4\left(\frac{1}{8}dd'\right)$ or $\frac{1}{2}dd'$.

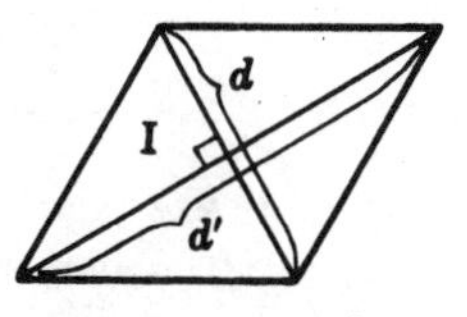

Rhombus: $A = \frac{1}{2}dd'$

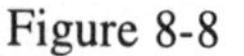

Figure 8-8

Polygons of the Same Size or Shape

Figure 8-9 shows what we mean when we say that two polygons are of equal area, or are similar, or are congruent.

Equal Polygons

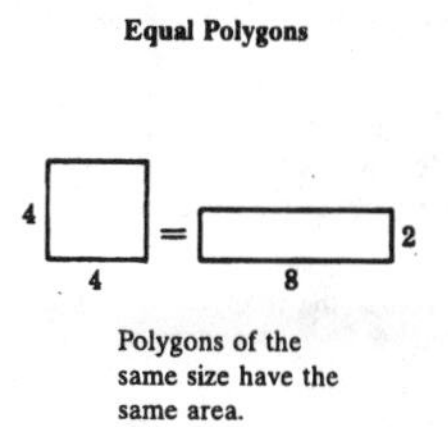

Polygons of the same size have the same area.

Similar Polygons

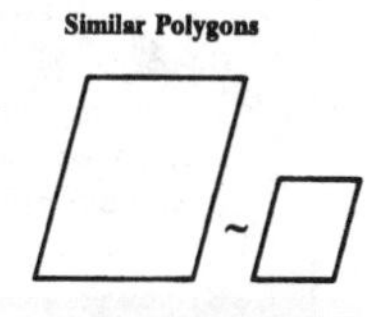

Similar polygons have the same shape.

Congruent Polygons

$\cong$

Congruent polygons have the same size and the same shape.

Figure 8-9

PRINCIPLE 1: *Parallelograms have equal areas if they have congruent bases and congruent altitudes.*

PRINCIPLE 2: *Triangles have equal areas if they have congruent bases and congruent altitudes.*

PRINCIPLE 3: *A median divides a triangle into two triangles with equal areas.*

PRINCIPLE 4: *Triangles are equal in area if they have a common base and their vertices lie on a line parallel to the base.*

Solved Problem 8-1. Prove that if M is the midpoint of diagonal $\overline{AC}$ in quadrilateral $ABCD$, and $\overline{BM}$ and $\overline{DM}$ are drawn, then the area of quadrilateral $ABMD$ equals the area of quadrilateral $CBMD$.

Solution

Given: Quadrilateral $ABCD$
M is midpoint of diagonal $\overline{AC}$.

To Prove: Area of quadrilateral $ABMD$ equals area of quadrilateral $CBMD$.

Plan: Use Principle 3 to obtain two pairs of triangles which are equal in area. Then use the Addition Postulate.

PROOF:

Statements	Reasons
1. M is the midpoint of $\overline{AC}$.	1. Given
2. $\overline{BM}$ is a median of $\triangle ACB$. $\overline{DM}$ is a median of $\triangle ACD$.	2. A line from a vertex of a triangle to the midpoint of the opposite side is a median.
3. Area ($\triangle AMB$) = area($\triangle BMC$), Area($\triangle AMD$) = area($\triangle DMC$).	3. A median divides a triangle into two triangles of equal area.
4. Area of quadrilateral $ABMD$ equals area of quadrilateral $CBMD$.	4. If equals are added to equals, the results are equal.

Chapter 9
Regular Polygons and the Circle

In This Chapter:

- ✔ *Regular Polygons*
- ✔ *Relationships of Segments in Regular Polygons of 3, 4, and 6 Sides*
- ✔ *Area of a Regular Polygon*
- ✔ *Ratios of Segments and Areas of Regular Polygons*
- ✔ *Circumference and Area of a Circle*
- ✔ *Length of an Arc; Area of a Sector and a Segment*
- ✔ *Areas of Combination Figures*

Regular Polygons

A *regular polygon* is an equilateral and equiangular polygon. The *center of a regular polygon* is the common center of its inscribed and circumscribed circles. A *radius of a regular polygon* is a segment joining its center to any vertex. A radius of a regular polygon is also a radius of the circumscribed circle. A *central angle of a regular polygon* is an angle included between two radii drawn to successive vertices. An *apothem of a regular polygon* is a segment from its center perpendicular to one of its sides. An apothem is also a radius of the inscribed circle.

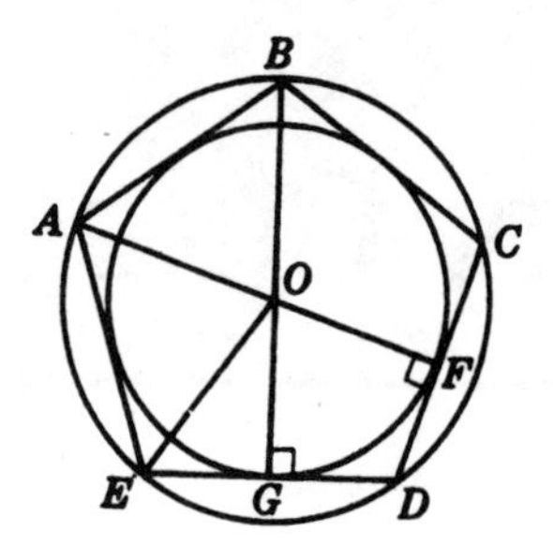

Figure 9-1

For the regular pentagon shown in Fig. 9-1, $\overline{AB} = \overline{BC} = \overline{CD} = \overline{DE} = \overline{EA}$ and $m\angle A = m\angle B = m\angle C = m\angle D = m\angle E$. Also, its center is O, $\overline{OA}$ and $\overline{OB}$ are its radii; $\angle AOB$ is a central angle; and $\overline{OG}$ and $\overline{OF}$ are apothems.

Regular-Polygon Principles

PRINCIPLE 1: *If a regular polygon of n sides has a side of length s, the perimeter is* $p = ns$.

PRINCIPLE 2: *A circle may be circumscribed about any regular polygon.*

PRINCIPLE 3: *A circle may be inscribed in any regular polygon.*

PRINCIPLE 4: *The center of the circumscribed circle of a regular polygon is also the center of its inscribed circle.*

PRINCIPLE 5: *An equilateral polygon inscribed in a circle is a regular polygon.*

Principle 6: *Radii of a regular polygon are congruent.*

Principle 7: *A radius of a regular polygon bisects the angle to which it is drawn.*

Thus, in Fig. 9-1, $\overline{OB}$ bisects $\angle ABC$.

Principle 8: *Apothems of a regular polygon are congruent.*

Principle 9: *An apothem of a regular polygon bisects the side to which it is drawn.*

Thus, in Fig. 9-1, $\overline{OF}$ bisects $\overline{CD}$, and $\overline{OG}$ bisects $\overline{ED}$.

Principle 10: *For a regular polygon of n sides:*

1. *Each central angle c measures* $\dfrac{360^o}{n}$.

2. *Each interior angle i measures* $\dfrac{(n-2)180^o}{n}$.

3. *Each exterior angle e measures* $\dfrac{360^o}{n}$.

Thus, for the regular pentagon ABCDE of Fig. 9-2,

$$m\angle AOB = m\angle ABS = \frac{360^o}{n} = \frac{360^o}{5} = 72^o$$

$$m\angle ABC = \frac{(n-2)180^o}{n} = \frac{(5-2)180^o}{5} = 108^o$$

and $m\angle ABC + m\angle ABS = 180°$

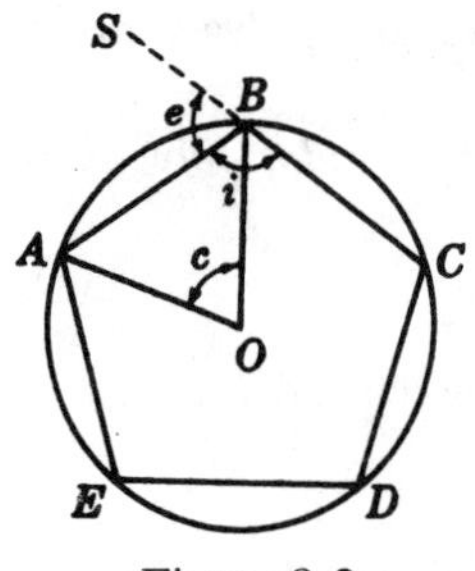

Figure 9-2

Relationships of Segments in Regular Polygons of 3, 4, and 6 Sides

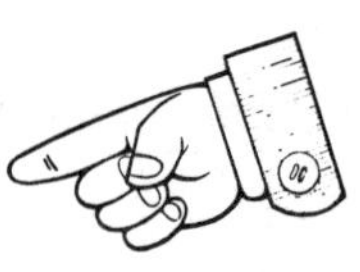

In the regular hexagon, square, and equilateral triangle, special right triangles are formed when the apothem r and a radius R terminating in the same side are drawn. In the case of the square, we obtain a 45°-45°-90° triangle, while in the other two cases, we obtain a 30°-60°-90° triangle. The formulas in Fig. 9-3 relate the lengths of the sides and radii of these regular polygons.

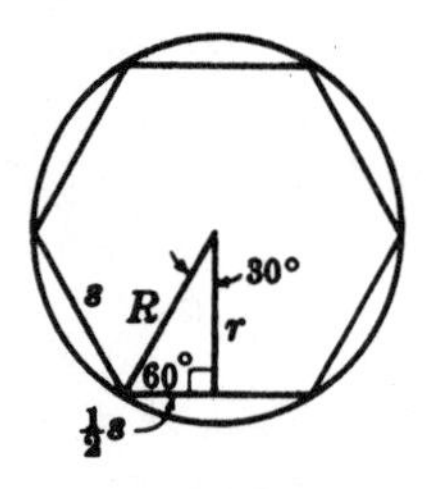

Regular Hexagon
$s = R$
$r = \frac{1}{2}R\sqrt{3}$

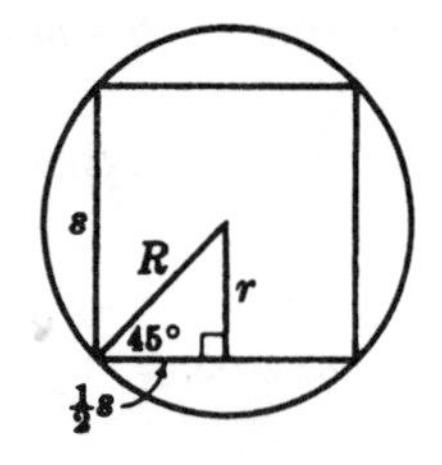

Square
$s = R\sqrt{2}$
$r = \frac{1}{2}s = \frac{1}{2}R\sqrt{2}$

Figure 9-3

Area of a Regular Polygon

The area of a regular polygon equals one-half the product of its perimeter and the length of its apothem. As shown in Fig. 9-4, by drawing radii, we can divide a regular polygon on n sides and perimeter $p = ns$ into n triangles, each of area $\frac{1}{2}rs$. Hence, the area of the regular polygon is $n\left(\frac{1}{2}rs\right) = \frac{1}{2}nsr = \frac{1}{2}pr$.

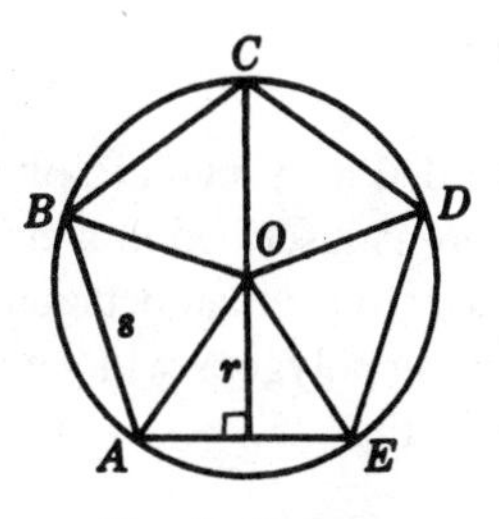

Regular Polygon
$A = \frac{1}{2}nsr = \frac{1}{2}pr$

Figure 9-4

Ratios of Segments and Areas of Regular Polygons

PRINCIPLE 1: *Regular polygons having the same number of sides are similar.*

PRINCIPLE 2: *Corresponding segments of regular polygons having the same number of sides are in proportion.* "Segments" here include sides, perimeters, radii or circumferences of circumscribed or inscribed circles, and such.

Principle 3: *Areas of regular polygons having the same number of sides are to each other as the squares of the lengths of any two corresponding segments.*

Circumference and Area of a Circle

π (pi) is the ratio of the circumference C of any circle to its diameter d; that is, $\pi = C/d$. Hence,

$C = \pi d$ or $C = 2\pi r$

Approximate value for π is 3.14.

A circle may be regarded as a regular polygon having an infinite number of sides. If a square is inscribed in a circle, and the number of sides is continually doubled (to form an octagon, a 16-gon, and so on), the perimeters of the resulting polygons will get closer and closer to the circumference of the circle (Fig 9-5).

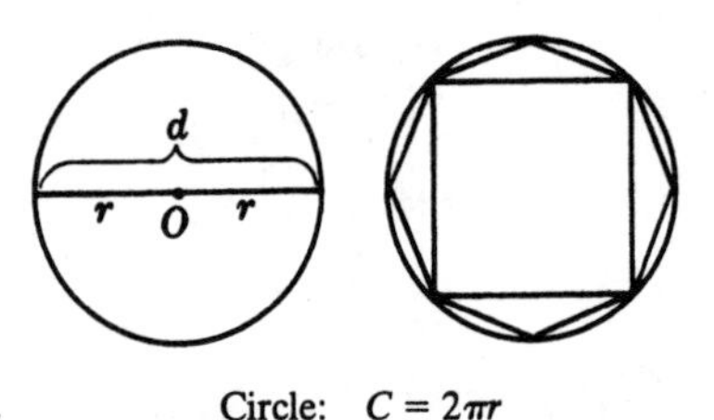

Circle: $C = 2\pi r$
$A = \pi r^2$

Figure 9-5

To find the area of a circle, the formula $A = \frac{1}{2}\pi r$ can be used with C substituted for p; doing so, we get $A = \frac{1}{2}Cr = \frac{1}{2}(2\pi r)(r) = \pi r^2$

All circles are similar figures since they have the same shape. Because they are similar figures, (1) corresponding segments of circles are in proportion and (2) the areas of two circles are to each other as the squares of their radii or circumferences.

Length of an Arc; Area of a Sector and a Segment

A sector of a circle is part of a circle bounded by two radii and their intercepted arc. Thus, in Fig. 9-6, the shaded area of circle O is a sector OAB.

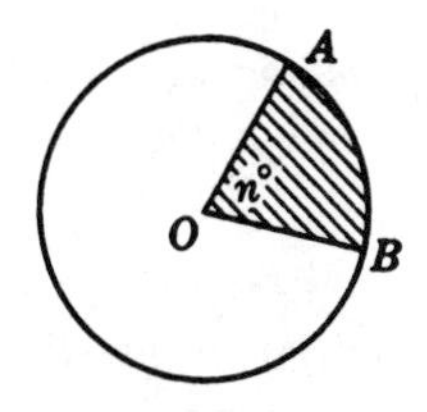

Figure 9-6

A *segment of a circle* is a part of a circle bounded by a chord and its arc. A *minor segment* of a circle is the smaller of the two segments thus formed. Thus, in Fig. 9-7, the shaded section of circle Q is the minor segment of $\widehat{ACB}$.

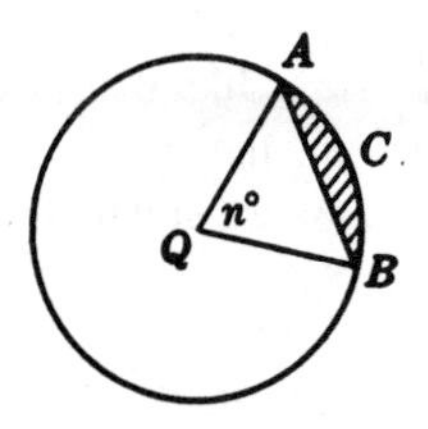

Figure 9-7

Principle 1: *In a circle of radius r, the length l of an arc of measure n° equals* $\frac{n}{360}$ *of the circumference of the circle, or*

$$l = \frac{n}{360} 2\pi r = \frac{\pi n r}{180}.$$

PRINCIPLE 2: *In a circle of radius r, the area K of a sector of measure $n°$ equals $\frac{n}{360}$ of the area of the circle, or* $K = \frac{n}{360}\pi r^2$.

PRINCIPLE 3:

$$\frac{\text{Area of a sector of n}^{\circ}}{\text{Area of the circle}} = \frac{\text{Length of an arc of measure n}^{\circ}}{\text{Circumference of the circle}} = \frac{n}{360}$$

PRINCIPLE 4: *The area of a minor segment of a circle equals the area of its sector less the area of the triangle formed by its radii and chord.*

PRINCIPLE 5: *If a regular polygon is inscribed in a circle, each segment cut off by the polygon has area equal to the difference between the area of the circle and the area of the polygon divided by the number of sides.*

Areas of Combination Figures

The areas of combination figures like that in Fig 9-8 may be found by determining individual areas and then adding or subtracting as required. Thus, the shaded area in the figure equals the sum of the areas of the square and the semicircle:

$$A = 8^2 + \frac{1}{2}(16\pi) = 64 + 8\pi$$

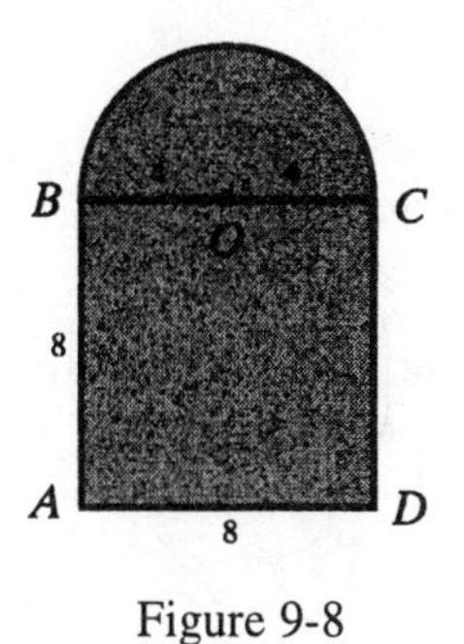

Figure 9-8

Solved Problem 9-1: Find the shaded area in each part of Fig. 9-9. In (a), circles A, B, and C are tangent externally and each has a radius 3. In (b), each arc is part of a circle of radius 9.

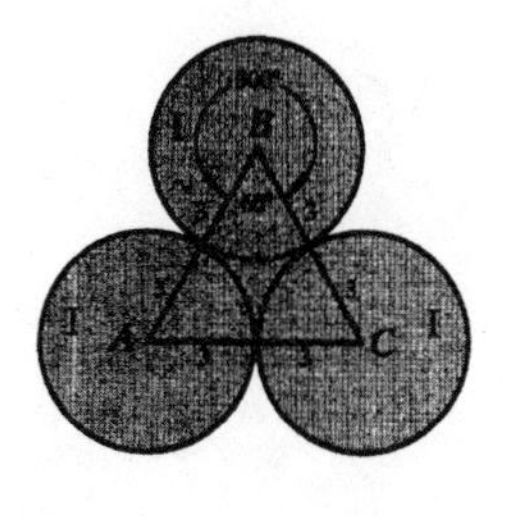

(a)

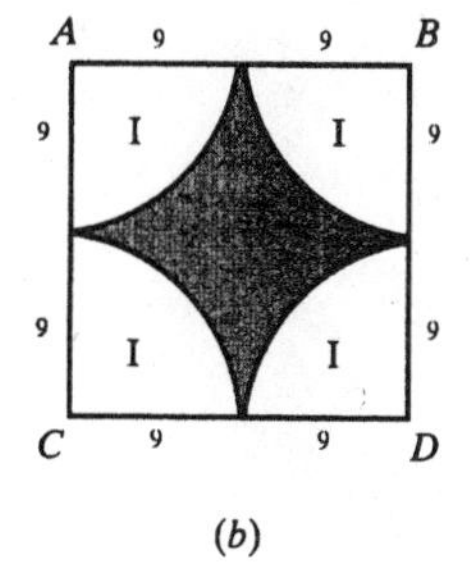

(b)

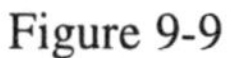

Figure 9-9

Solution:

(a) Area of $\Delta ABC = \frac{1}{4}s^2\sqrt{3} = \frac{1}{4}(6^2)\sqrt{3} = 9\sqrt{3}$

$$\text{Area of sector I} = \frac{n^o}{360^o}(\pi r^2) = \frac{300}{360}(9\pi) = \frac{15}{2}\pi$$

$$\text{Shaded area} = 9\sqrt{3} + 3\left(\frac{15}{2}\pi\right) = 9\sqrt{3} + \frac{45}{2}\pi$$

(b) Area of square = $18^2 = 324$.

Area of sector I = $\frac{n^o}{360^o}(\pi r^2) = \frac{90}{360}(81\pi) = \frac{81}{4}\pi$

Shaded area = $324 - 4\left(\frac{81}{4}\pi\right) = 324 - 81\pi$

Chapter 10
CONSTRUCTIONS

IN THIS CHAPTER:

- ✔ *Duplicating Segments and Angles*
- ✔ *Constructing Bisectors and Perpendiculars*
- ✔ *Constructing a Triangle*
- ✔ *Constructing Parallel Lines*
- ✔ *Circle Constructions*
- ✔ *Inscribing and Circumscribing Regular Polygons*
- ✔ *Constructing Similar Triangles*

Duplicating Segments and Angles

CONSTRUCTION 1: *To construct a line segment congruent to a given line segment*

Given: Line segment $\overline{AB}$ (Fig. 10-1).

To construct: A line segment congruent to $\overline{AB}$

Construction: On a working line w, with any point C as a center and a radius equal to $\overline{AB}$, construct an arc intersecting w at D. Then $\overline{CD}$ is the required line segment.

Figure 10-1

Solved Problem 10-1: Given line segments with lengths a and b (Fig. 10-2), construct line segments with lengths equal to (a) $a + 2b$; (b) $2(a + b)$; and (c) $b - a$.

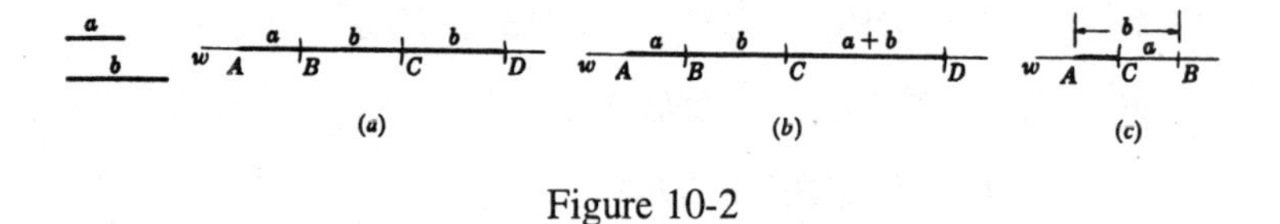

Figure 10-2

Solution: Using Construction 1:

(a) On a working line w, construct a line segment $\overline{AB}$ with length a. From B, construct a line segment with length equal to b, to point C; and from C, construct a line segment with length b, to point D. Then $\overline{AD}$ is the required line segment.

(b) Similar to (a), $AD = a + b + (a + b)$.

(c) Similar to (a), first construct $\overline{AB}$ with length b, then $\overline{BC}$ with length a. $AC = b - a$.

CONSTRUCTION 2: *To construct an angle congruent to a given angle*
Given: $\angle A$ (Fig. 10-3).

To construct: An angle congruent to $\angle A$.
Construction: With A as center and a convenient radius, construct an arc (1) intersecting the sides of $\angle A$ at B and C. With A', a point on a working line w, as center and the same radius, construct arc (2) inter-

secting w at B'. With B' as center and a radius equal to $\overline{BC}$, construct arc (3) intersecting arc (2) at C'. Draw $\overline{A'C'}$. Then $\angle A'$ is the required angle.

$(\triangle ABC \cong \triangle ADC$ by s.s.s. $\cong$ s.s.s.; hence $\angle 1 \cong \angle 2)$

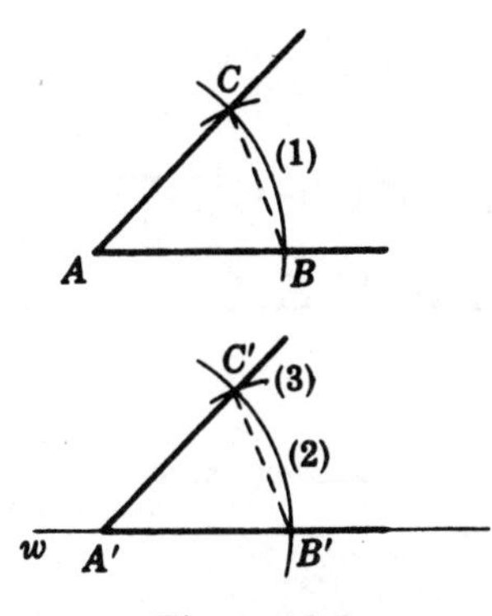

Figure 10-3

Solved Problem 10-2: Given $\triangle ABC$ in Fig. 10-4, construct angles whose measures are equal to (a) $2A$; (b) $A + B + C$; and (c) $B - A$.

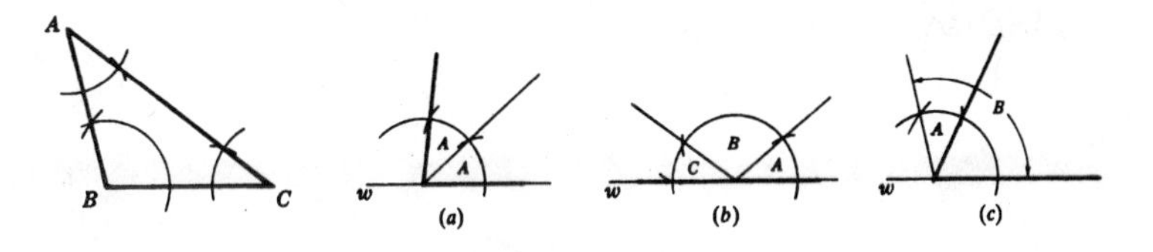

Figure 10-4

Solution: Using Construction 2:

(a) Using a working line w as one side, duplicate $\angle A$. Construct another duplicate of $\angle A$ adjacent to $\angle A$, as shown. The exterior sides of the copied angles form the required angle.

(b) Using a working line w as one side, duplicate $\angle A$. Construct $\angle B$ adjacent to $\angle A$. Then construct $\angle C$ adjacent to $\angle B$. The exterior sides of the copied angles A and C form the required angle. Note that the angle is a straight angle.

(c) Using a working line w as one side, duplicate $\angle B$. Then duplicate $\angle A$ from the new side of $\angle B$ as shown. The difference is the required angle.

Constructing Bisectors and Perpendiculars

Construction 3: *To bisect a given angle*

Given: $\angle A$ (Fig. 10-5).

To construct: The bisector of $\angle A$.

Construction: With A as center and a convenient radius, construct an arc intersecting the sides of $\angle A$ at B and C. With B and C as centers and equal radii, construct arcs intersecting in D. Draw $\overrightarrow{AD}$. Then $\overrightarrow{AD}$ is the required bisector.

$(\triangle ABD \cong \triangle ADC$ by s.s.s. $\cong$ s.s.s.; hence $\angle 1 \cong \angle 2)$

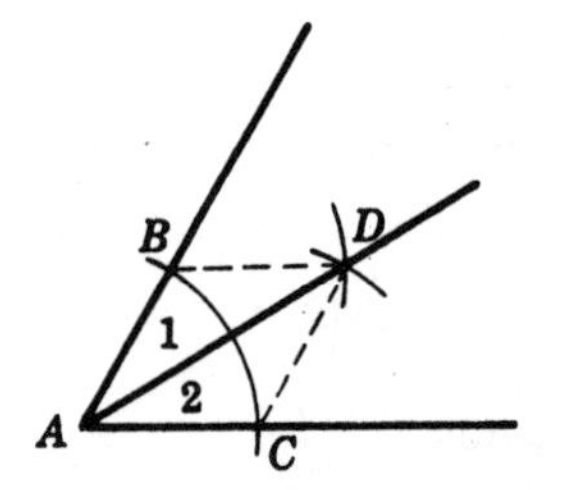

Figure 10-5

Construction 4: *To construct a line perpendicular to a given line through a given point on the line*

Given: Line w and point P on w (Fig. 10-6).
To construct: A perpendicular to w at P.
Construction: Using Construction 3, bisect the straight angle at P. Then $\overrightarrow{DP}$ is the required perpendicular; $\overleftrightarrow{DP}$ is the required line.

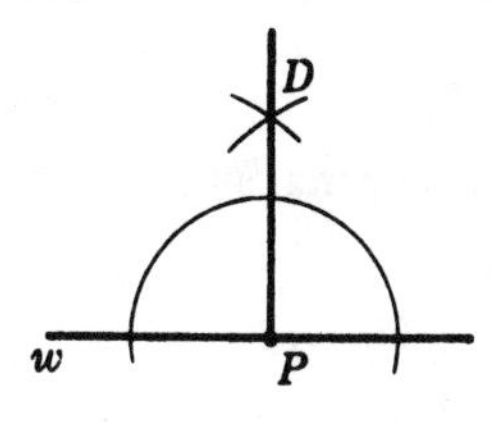

Figure 10-6

Construction 5: *To bisect a given line segment (to construct the perpendicular bisector of a given line segment)*

Given: Line segment $\overline{AB}$ (Fig. 10-7).

To construct: The perpendicular bisector of $\overline{AB}$.

Construction: With A as center and a radius of more than half $\overline{AB}$, construct arc (1). With B as center and the same radius, construct arc (2) intersecting arc (1) at C and D. Draw $\overleftrightarrow{CD}$. $\overleftrightarrow{CD}$ is the required perpendicular bisector of $\overline{AB}$. (Two points each equidistant from the ends of a segment determine the perpendicular bisector of the segment.)

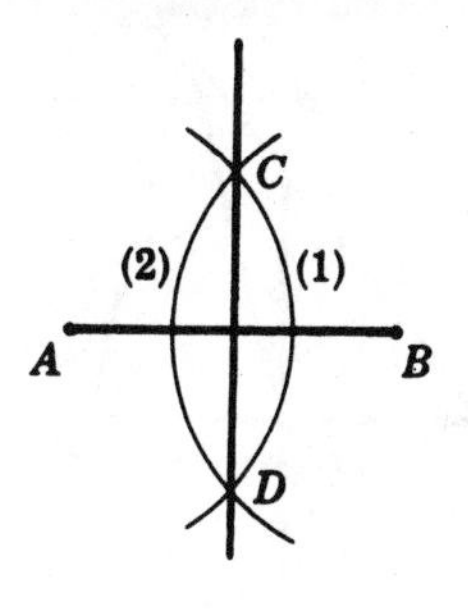

Figure 10-7

CONSTRUCTION 6: *To construct a line perpendicular to a given line through a given external point*
Given: Line w and point P outside of w (Fig. 10-8).
To construct: A perpendicular to w through P.
Construction: With P as center and a sufficiently long radius, construct an arc intersecting w at B and C. With B and C as centers and equal radii of more than half $\overline{BC}$, construct arcs intersecting at A. Draw $\overleftrightarrow{PA}$. Then $\overleftrightarrow{PA}$ is the required perpendicular. (Points P and A are each equidistant from B and C.)

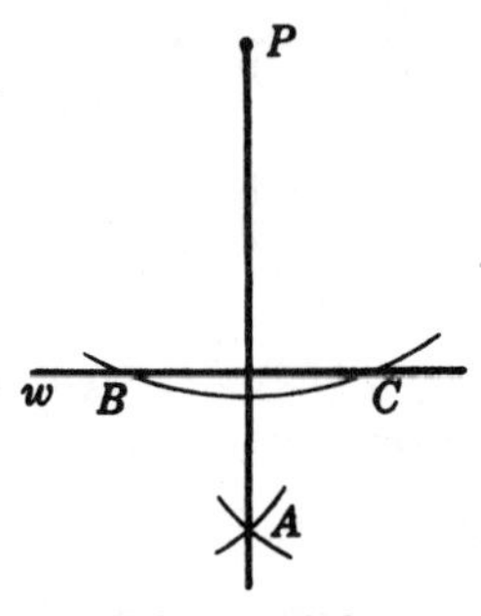

Figure 10-8

Solved Problem 10-3: In scalene $\triangle ABC$ [Fig. 10-9(a)], construct (a) a perpendicular bisector of $\overline{AB}$ and (b) a median to $\overline{AB}$. In $\triangle DEF$ [Fig. 10-9(b)], D is an obtuse angle; construct (c) the altitude to $\overline{DF}$ and (d) the bisector of $\angle E$.

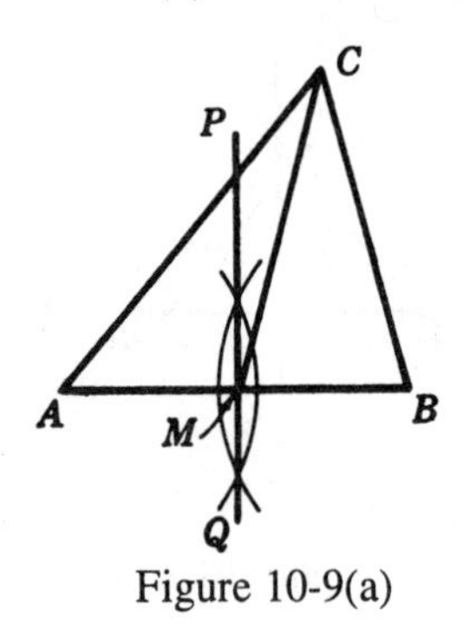

Figure 10-9(a)

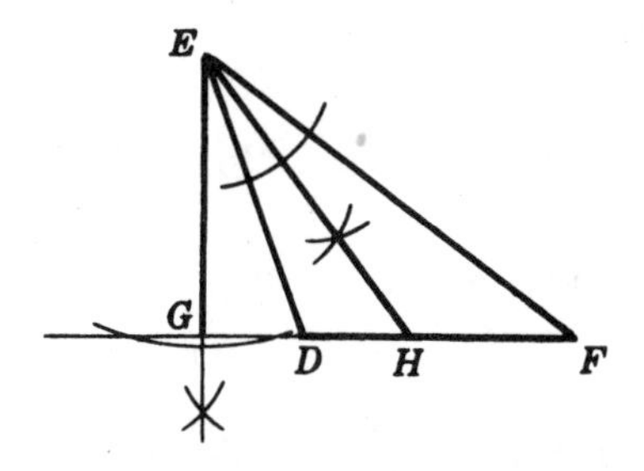

Figure 10-9(b)

Solution:

(a) Use Construction 5 to obtain $\overleftrightarrow{PQ}$, the perpendicular bisector of $\overline{AB}$.

(b) Point M is the midpoint of $\overline{AB}$. Draw $\overline{CM}$, the median to $\overline{AB}$.

(c) Use Construction 6 to obtain $\overline{EG}$, the altitude to $\overline{DF}$ (extended).

(d) Use Construction 3 to bisect $\angle E$. $\overrightarrow{EH}$ is the required bisector.

Solved Problem 10-4: (a) Construct angles measuring 90°, 45°, and 135°. (b) Given an angle with measure A (Fig. 10-10), construct an angle whose measure is $90° + A$.

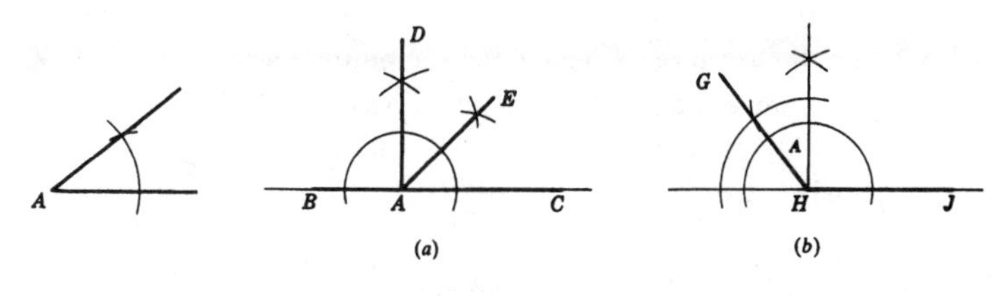

Figure 10-10

Solution:

(a) In Fig. 10-10(a), $m\angle DAB = 90°$, $m\angle CAE = 45°$, $m\angle BAE = 135°$.

(b) In Fig. 10-10(b), $m\angle GHJ = 90° + A$.

Constructing a Triangle

Determining a Triangle

A triangle is determined when a set of given data fix its size and shape. Since the parts needed to prove congruent triangles fix the size and shape of the triangles, a triangle is determined when the given data consist of three sides, or two sides and the angle included by those sides, or two angles and a side included by those angles, or two angles and a side not included by those angles, or the hypotenuse and either leg of a right triangle.

Sketching Triangles to be Constructed

Before doing the actual construction, it is very helpful to make a preliminary sketch of the required triangle. In this sketch:

1. Show the position of each of the given parts of the triangle.
2. Draw the given parts heavy, the remaining parts light.
3. Approximate the sizes of the given parts.
4. Use small letters for sides to agree with the capital letters for the angles opposite them.

As an example, you might make a sketch like that in Fig. 10-11 before constructing a triangle given two angles and an included side.

Given Parts

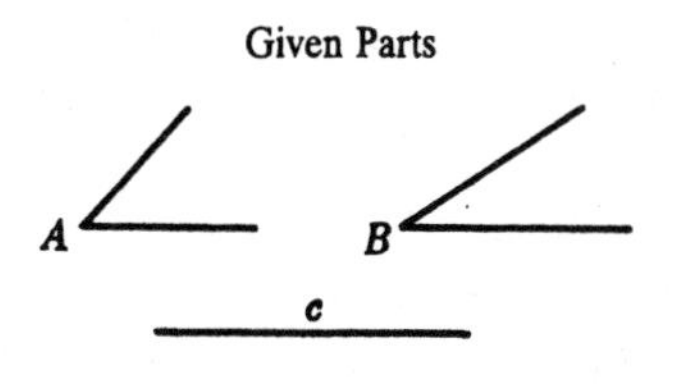

Preliminary Sketch

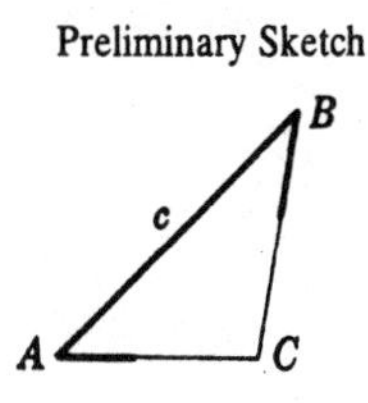

Figure 10-11

Triangle Constructions

Construction 7: *To construct a triangle given its three sides*
Given: Sides of lengths a, b, and c (Fig. 10-12).

To construct: $\triangle ABC$

Construction: On a working line w, construct $\overline{AC}$ such that $AC = b$. With A as center and c as radius, construct arc (1). Then with C as center and a as radius, construct arc (2) intersecting arc (1) at B. Draw $\overline{BC}$ and $\overline{AB}$. $\triangle ABC$ is the required triangle.

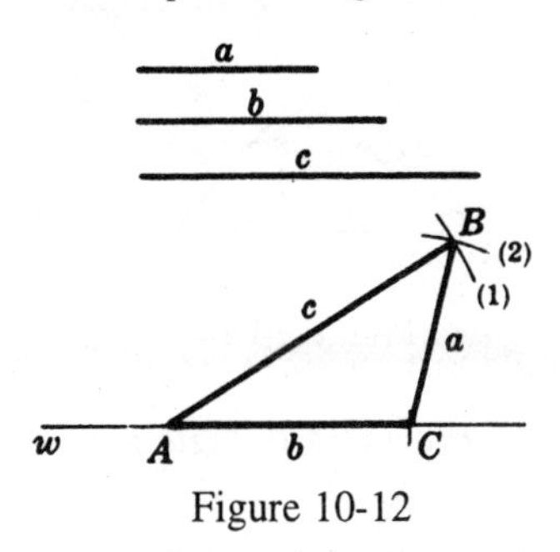

Figure 10-12

Construction 8: *To construct an angle of measure 60°*
Given: Line w (Fig. 10-13).
To construct: An angle of measure 60°
Construction: Using a convenient length as a side, construct an equilateral triangle using Construction 7. Then any angle of the equilateral triangle is the required angle.

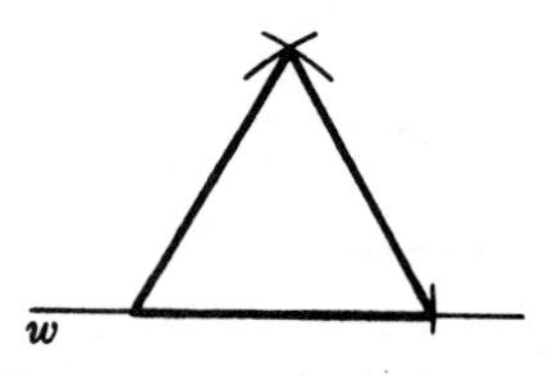

Figure 10-13

CONSTRUCTION 9: *To construct a triangle given two sides and the included angle*

Given: $\angle A$, segments of lengths b and c (Fig. 10-14).

To construct: $\triangle ABC$

Construction: On a working line w, construct $\overline{AC}$ such that $AC = b$. At A, construct $\angle A$ with one side on $\overline{AC}$. On the other side of $\angle A$, construct $\overline{AB}$ such that $AB = c$. Draw $\overline{BC}$. Then the required triangle is $\triangle ABC$.

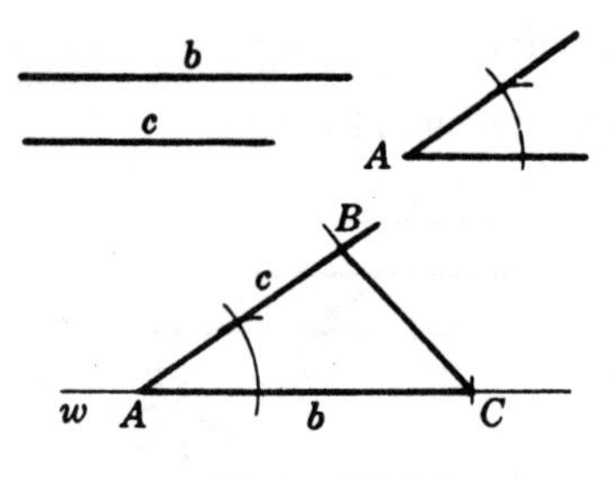

Figure 10-14

CONSTRUCTION 10: *To construct a triangle given two angles and the included side*

Given: $\angle A$, $\angle C$, and a segment of length b (Fig. 10-15).

To construct: $\triangle ABC$

Construction: On a working line w, construct $\overline{AC}$ such that $AC = b$. At A, construct $\angle A$ with one side on $\overline{AC}$ and at C construct $\angle C$ with one side on $\overline{AC}$. Extend the new sides of the angles until they meet, at B.

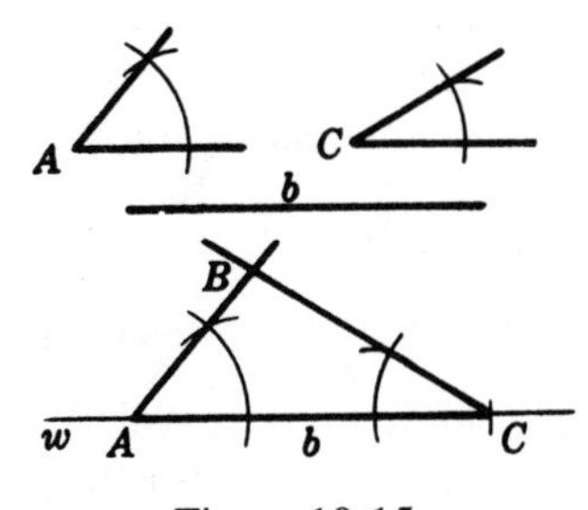

Figure 10-15

CONSTRUCTION 11: *To construct a triangle given two angles and a side not included*

Given: $\angle A$, $\angle C$, and a segment of length b (Fig. 10-16).

To construct: $\triangle ABC$

Construction: On a working line w, construct $\overline{AC}$ such that $\overline{AC} = b$. At C, construct an angle with measure equal to $m\angle A + m\angle B$ so that the extension of $\overline{AC}$ will be one side of the angle. The remainder of the straight angle at C will be $\angle C$. At A, construct $\angle A$ with one side on $\overline{AC}$. The intersection of the new sides of the angles is B.

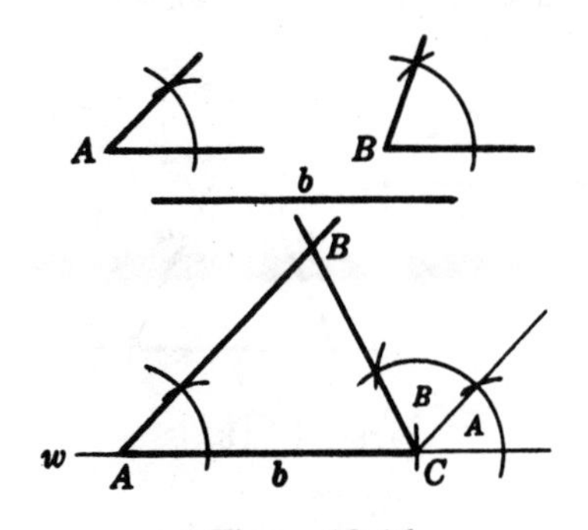

Figure 10-16

CONSTRUCTION 12: *To construct a right triangle given its hypotenuse and a leg*
Given: Hypotenuse with length c and leg with length b of right triangle ABC (Fig. 10-17).
To construct: Right triangle ABC

Construction: On a working line w, construct $\overline{AC}$ such that $\overline{AC} = b$. At C, construct a perpendicular to $\overline{AC}$. With A as center and a radius of c, construct an arc intersecting the perpendicular at B.

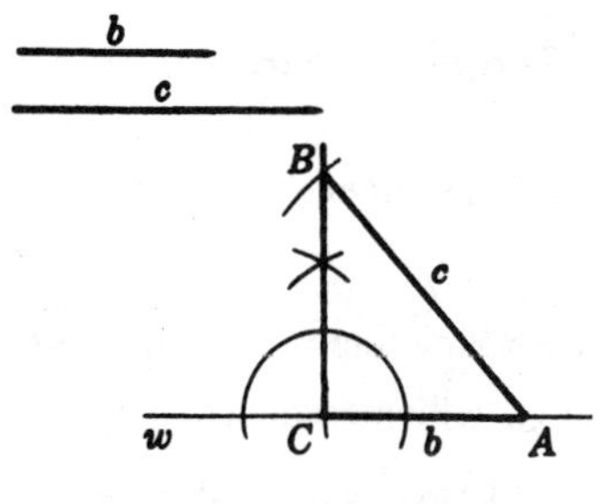

Figure 10-17

Solved Problem 10-5: Construct an isosceles triangle, given the lengths of the base and an arm (Fig. 10-18).

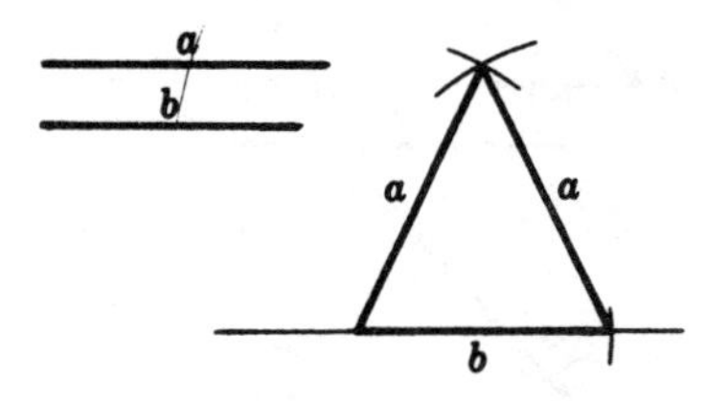

Figure 10-18

Solution: Use Construction 7, since all three sides of the triangle are known.

Solved Problem 10-6: Construct an angle of measure (a) 120°; (b) 30°; (c) 150°; (d) 105°; and (e) 75°.

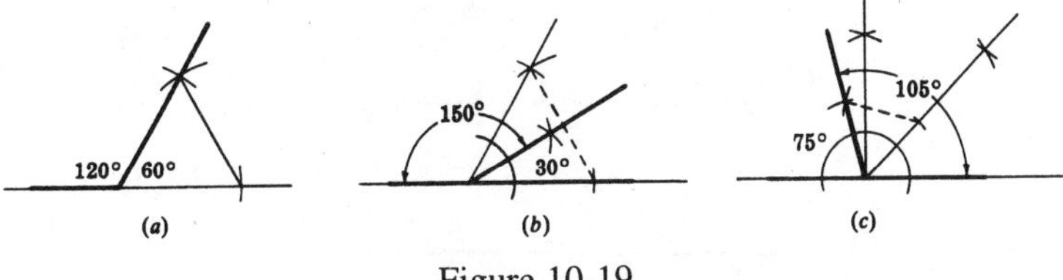

Figure 10-19

Solution:

(a) Use Construction 8 [Fig. 10-19(a)] to construct 120° as 180° – 60°.

(b) Use Constructions 8 and 3 to construct 30° as $\frac{1}{2}(60^\circ)$ [Fig. 10-19(b)].

(c) Use (b) to construct 150° as 180° – 30° [Fig. 10-19(b)].

(d) Use Constructions 3, 4, and 8 to construct 105° as $60^\circ + \frac{1}{2}(90^\circ)$ [Fig. 10-19(c)].

(e) Use (d) to construct 75° as 180° – 105° [Fig. 10-19(c)].

Constructing Parallel Lines

CONSTRUCTION 13: *To construct a line parallel to a given line through a given external point*

Given: $\overleftrightarrow{AB}$ and external point P (Fig. 10-20).

To construct: A line through P parallel to $\overleftrightarrow{AB}$.

Construction: Draw a line $\overleftrightarrow{RS}$ through P intersecting $\overleftrightarrow{AB}$ in Q. Construct $\angle SPD \cong \angle PQB$. Then, $\overleftrightarrow{CD}$ is the required parallel. (If two corresponding angles are congruent, the lines cut by the transversal are parallel.)

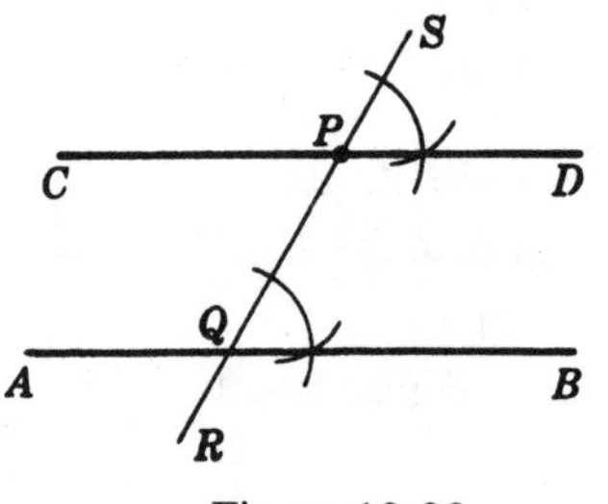

Figure 10-20

Solved Problem 10-7: Construct a parallelogram given the lengths of two adjacent sides a and b and of a diagonal d (Fig. 10-21).

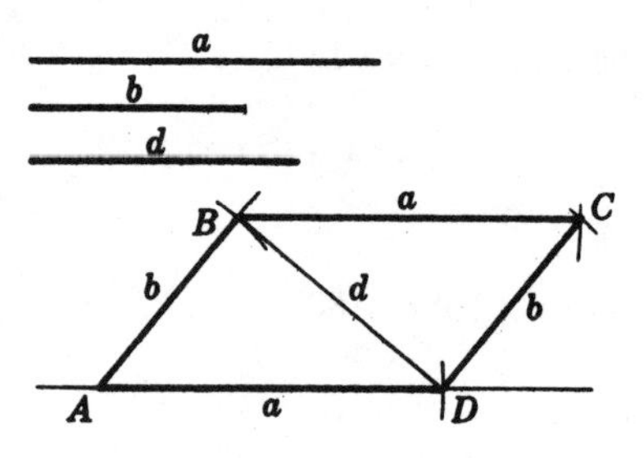

Figure 10-21

Solution:

Three vertices of the parallelogram are obtained by constructing $\triangle ABD$ by Construction 7. The fourth vertex, C, is obtained by constructing $\triangle BCD$ upon diagonal $\overline{BD}$ by Construction 7. Vertex C may also be obtained by constructing $\overline{BC} \parallel \overline{AD}$ and $\overline{DC} \parallel \overline{AB}$.

Circle Constructions

Construction 14: *To construct a tangent to a given circle through a given point on the circle*

Given: Circle O and point P on the circle (Fig. 10-22).
To construct: A tangent to circle O at P.

Construction: Draw radius $\overline{OP}$ and extend it outside the circle. Construct $\overleftrightarrow{AB} \perp \overleftrightarrow{OP}$ at P. $\overleftrightarrow{AB}$ is the required tangent. (A line perpendicular to a radius at its outer extremity is a tangent to the circle.)

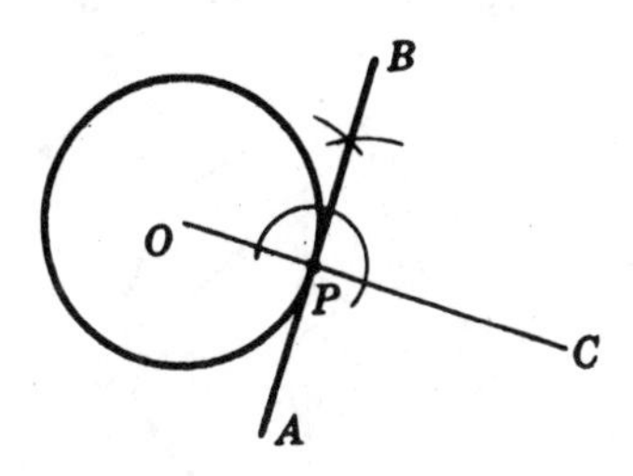

Figure 10-22

Construction 15: *To construct a tangent to a given circle through a given point outside the circle*
Given: Circle O and point P outside the circle (Fig. 10-23).
To construct: A tangent to circle O at P.

Construction: Draw $\overline{OP}$, and make $\overline{OP}$ the diameter of a new circle Q. Connect P to A and B, the intersections of circles O and Q. Then $\overline{PA}$ and $\overline{PB}$ are tangents. ($\angle OAP$ and $\angle OBP$ are right angles, since angles inscribed in semicircles are right angles.)

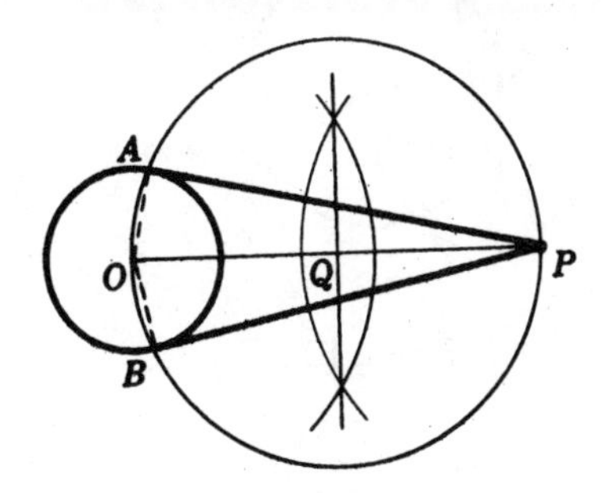

Figure 10-23

Construction 16: *To circumscribe a circle about a triangle*

Given: $\triangle ABC$ (Fig. 10-24).

To construct: The circumscribed circle of $\triangle ABC$.
Construction: Construct the perpendicular bisectors of two sides of the triangle. Their intersection is the center of the required circle, and the diameter to any vertex is the radius. (Any point on the perpendicular bisector of a segment is equidistant from the ends of a segment.)

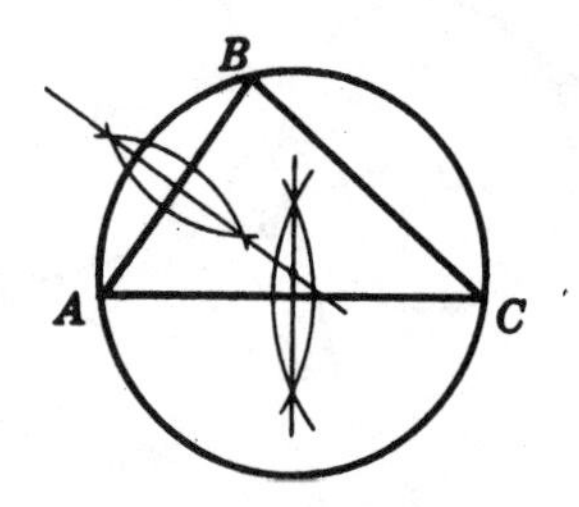

Figure 10-24

Construction 17: *To locate the center of a given circle*
Given: A circle (Fig. 10-25).
To construct: The center of the given circle.
Construction: Select any three points A, B, and C on the circle. Construct the perpendicular bisectors of line segments $\overline{AB}$ and $\overline{AC}$. The intersection of these perpendicular bisectors is the center of the circle.

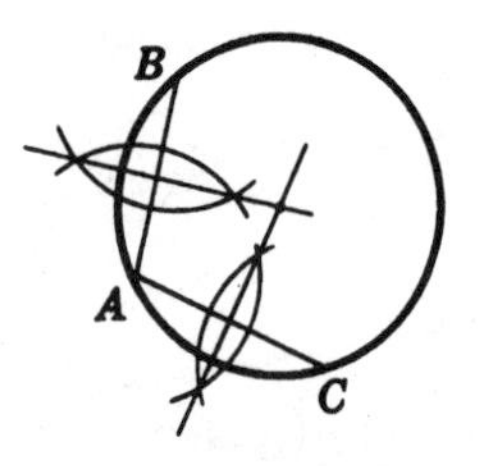

Figure 10-25

CONSTRUCTION 18: *To inscribe a circle in a given triangle*

Given: $\triangle ABC$ (Fig. 10-26).

To construct: The circle inscribed in $\triangle ABC$.

Construction: Construct the bisectors of two of the angles of $\triangle ABC$. Their intersection is the center of the required circle, and the distance (perpendicular) to any side is the radius. (Any point on the bisector of an angle is equidistant from the sides of the angle.)

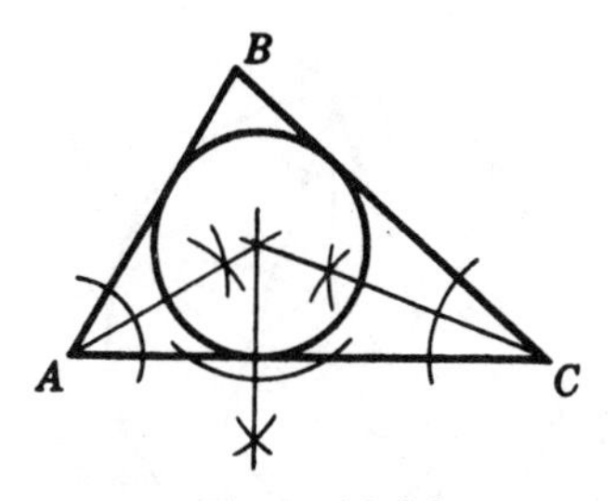

Figure 10-26

Solved Problem 10-8: A secant from a point P outside circle O in Fig. 10-27 meets the circle in B and A. Construct a triangle circumscribed about the circle so that two of its sides meet in P and the third side is tangent to the circle at A.

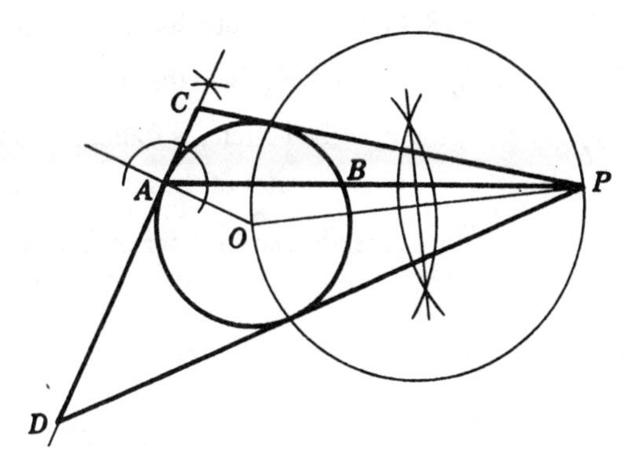

Figure 10-27

Solution:

Use Constructions 14 and 15: At A, construct a tangent to circle O. From P, construct tangents to circle O intersecting the first tangent in C and D. The required triangle is $\triangle PCD$.

Solved Problem 10-9: Construct the circumscribed and inscribed circles of isosceles triangle DEF in Fig. 10-28.

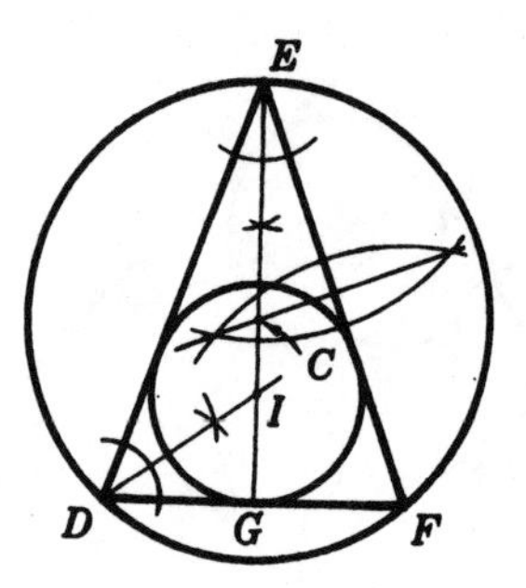

Figure 10-28

Solution:

Use Constructions 16 and 18. In doing so, note that the bisector of $\angle E$ is also the perpendicular bisector of $\overline{DF}$. Then the center of each circle is on $\overline{EG}$. I, the center of the inscribed circle, is found by constructing the bisector of $\angle D$ or $\angle F$. C, the center of the circumscribed circle, is found by constructing the perpendicular bisector of $\overline{DE}$ or $\overline{EF}$.

Inscribing and Circumscribing Regular Polygons

CONSTRUCTION 19: *To inscribe a square in a given circle*
Given: Circle O (Fig. 10-29).
To construct: A square inscribed in circle O.
Construction: Draw a diameter, and construct another diameter perpendicular to it. Join the endpoints of the diameters to form the required square.

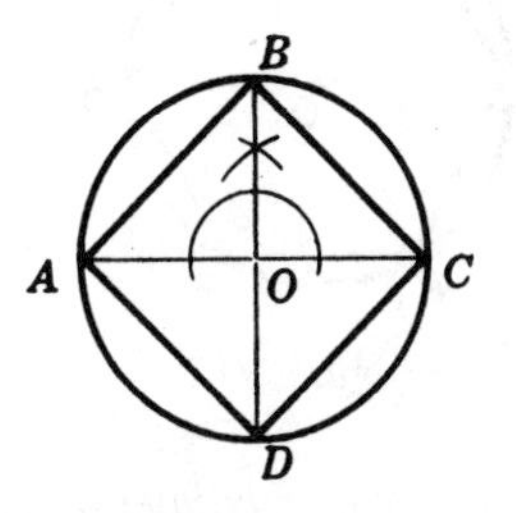

Figure 10-29

CONSTRUCTION 20: *To inscribe a regular octagon in a given circle*
Given: Circle O (Fig. 10-30).
To construct: A regular octagon inscribed in circle O.
Construction: As in Construction 19, construct perpendicular diameters. Then bisect the angles formed by these diameters, dividing the circle into eight congruent arcs. The chords of these arcs are the sides of the required regular octagon.

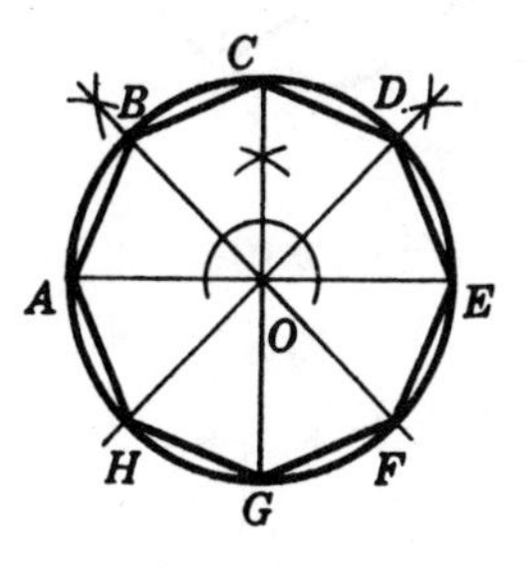

Figure 10-30

Construction 21: *To inscribe a regular hexagon in a given circle*
Given: Circle O (Fig. 10-31).
To construct: A regular hexagon inscribed in circle O.

Construction: Draw diameter $\overline{AD}$ and, using A and D as centers, construct four arcs having the same radius as circle O and intersecting the circle. Construct the required regular hexagon by joining consecutive points in which these arcs intersect the circle.

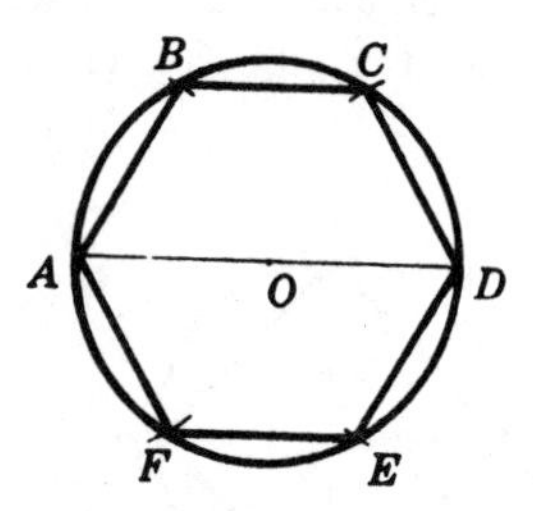

Figure 10-31

Construction 22: *To inscribe an equilateral triangle in a given circle*
Given: Circle O (Fig. 10-32).
To construct: An equilateral triangle inscribed in circle O.
Construction: Inscribed equilateral triangles are obtained by joining alternately the six points of division obtained in Construction 21.

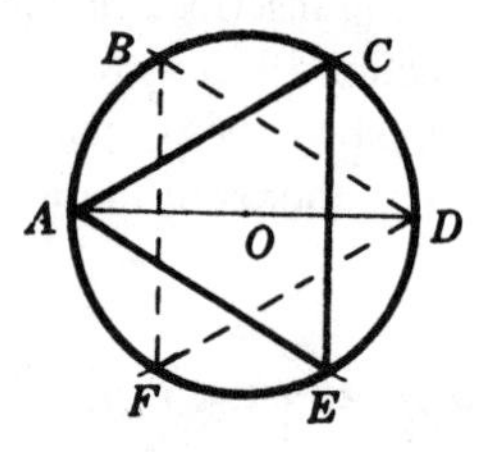

Figure 10-32

Constructing Similar Triangles

CONSTRUCTION 23: *To construct a triangle similar to a given triangle on a given line segment as base*

Given: Given $\triangle ABC$ and line segment $\overline{A'C'}$ (Fig. 10-33).

To construct: $\triangle A'B'C' \sim \triangle ABC$ on $\overline{A'C'}$ as base

Construction: On $\overline{A'C'}$, construct $\angle A' \cong \angle A$ and $\angle C' \cong \angle C$ using Construction 2. Extend the other sides until they meet, at B. (If two angles of one triangle are congruent to two angles of another triangle, the triangles are similar.)

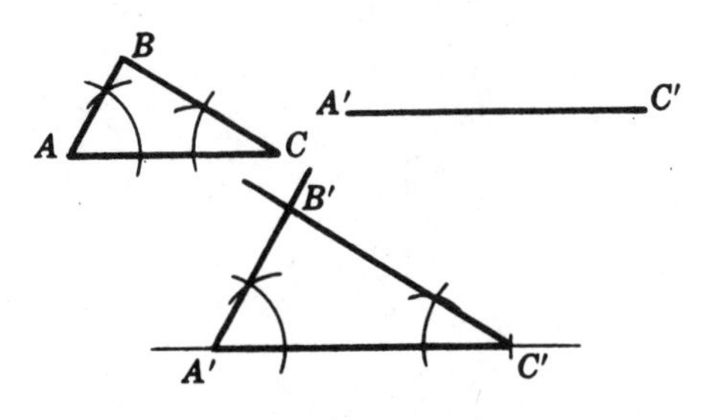

Figure 10-33

Solved Problem 10-10: Construct a triangle similar to triangle ABC in Fig. 10-34, with a base twice as long as the base of the given triangle.

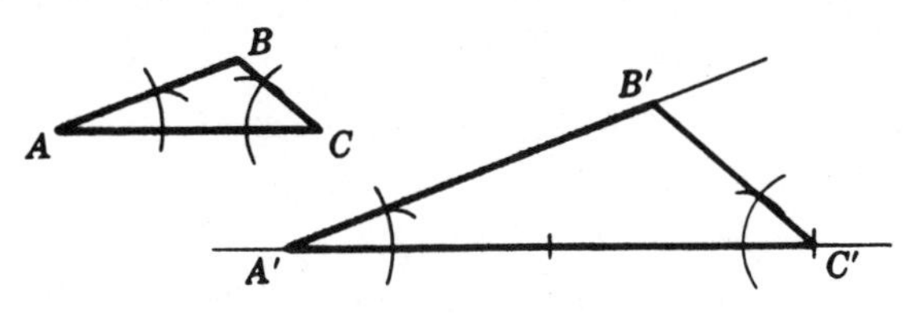

Figure 10-34

Solution:

Construct $\overline{A'C'}$ twice as long as $\overline{AC}$, and then use Construction 23.

Alternative method (Fig. 10-35): Extend two sides of $\triangle ABC$ to twice their lengths and join the endpoints.

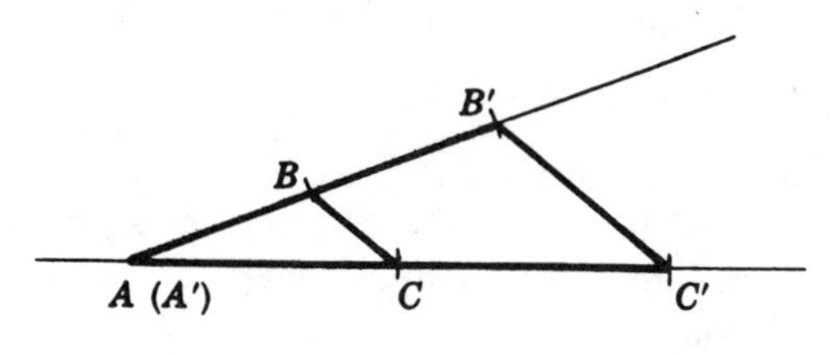

Figure 10-35

Appendix A
FORMULAS FOR REFERENCE

Angle Formulas

1. Complement of $a°$	1. $c = 90° - a°$
2. Supplement of $a°$	2. $s = 180 - a°$
3. Sum of measures of angles of a triangle	3. $S = 180°$
4. Sum of measures of angles of a quadrilateral	4. $S = 360°$
5. Sum of measures of exterior angles of an n-gon	5. $S = 360°$
6. Sum of measures of interior angles of an n-gon	6. $S = 180°(n - 2)$
7. Measure of each interior angle of an equiangular or regular n-gon	7. $S = \frac{180°(n-2)}{n}$
8. Measure of each exterior angle of an equiangular or regular n-gon	8. $S = \frac{360°}{n}$
9. Measure of central $\angle O$ intercepting an arc of $a°$	9. $m\angle O = a°$
10. Measure of inscribed $\angle A$ intercepting an arc of $a°$	10. $m\angle A = - \frac{1}{2}a°$
11. Measure of $\angle A$ formed by a tangent and a chord and intercepting an arc of $a°$	11. $m\angle A = - \frac{1}{2}a°$
12. Measure of $\angle A$ formed by two intersecting chords and intercepting arcs of $a°$ and $b°$	12. $m\angle A = \frac{1}{2}(a° + b°)$
13. Measure of $\angle$A formed by two intersecting tangents, two intersecting secants, or an intersecting tangent and secant and intercepting arcs of $a°$ and $b°$	13. $m\angle A = \frac{1}{2}(a° + b°)$
14. Measure of $\angle A$ inscribed in a semicircle	14. $m\angle A=90°$
15. Measure of $\angle$s and B of an inscribed quadrilateral	15. $m\angle A =180° - m\angle B$

Area Formulas

1. Area of a rectangle	1. $K = bh$
2. Area of a square	2. $K = s^2$, $\quad K = \frac{1}{2}d^2$
3. Area of a parallelogram	3. $K = bh$, $\quad K = ab \sin C$
4. Area of a triangle	4. $K = \frac{1}{2}bh$, $\quad K = \frac{1}{2}ab \sin C$
5. Area of a trapezoid	5. $K = \frac{1}{2}h(b + b')$, $\quad K = hm$
6. Area of an equilateral triangle	6. $K = \frac{1}{4}s^2\sqrt{3}$, $\quad K = \frac{1}{3}h^2\sqrt{3}$
7. Area of a rhombus	7. $K = \frac{1}{2}dd'$
8. Area of a regular polygon	8. $K = \frac{1}{2}pr$
9. Area of a circle	9. $K = \pi r^2$, $\quad K = \frac{1}{4}\pi d^2$
10. Area of a sector	10. $K = \frac{n}{360}(\pi r^2)$
11. Area of a minor segment	11. K = area of sector − area of triangle

Circle Intersection Formulas

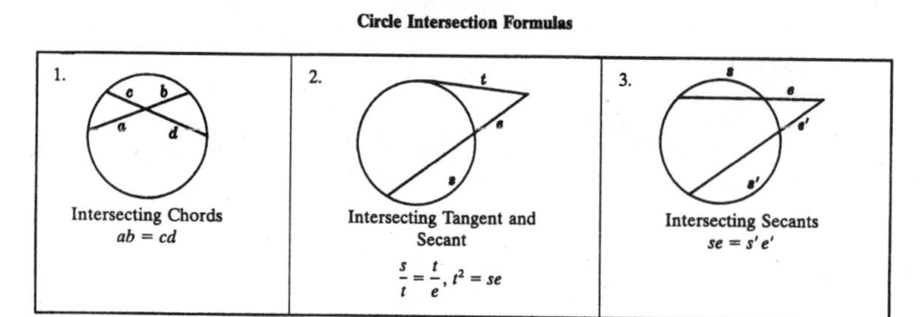

1. 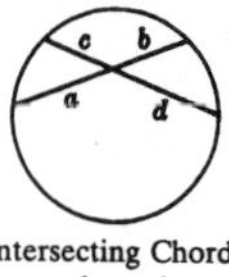 Intersecting Chords $ab = cd$	2. 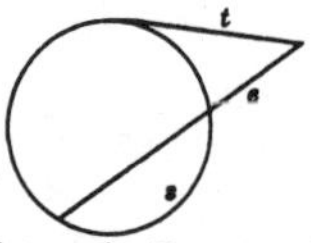 Intersecting Tangent and Secant $\frac{s}{t} = \frac{t}{e}$, $t^2 = se$	3. 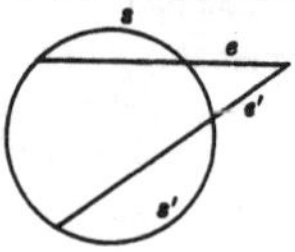 Intersecting Secants $se = s'e'$

Right-Triangle Formulas

1. 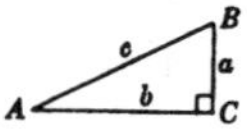	Pythagorean Theorem	1. $c^2 = a^2 + b^2$
2. 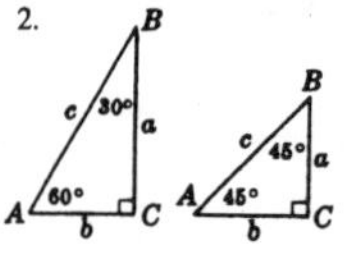	Leg opposite 30° angle Leg opposite 45° angle Leg opposite 60° angle	2. $b = \frac{1}{2}c$ $b = \frac{1}{2}c\sqrt{2}$, $b = a$ $a = \frac{1}{2}c\sqrt{3}$, $a = b\sqrt{3}$
3. 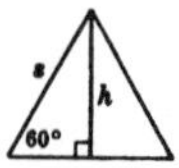	Altitude of equilateral triangle Side of equilateral triangle	3. $h = \frac{1}{2}s\sqrt{3}$ $s = \frac{2}{3}h\sqrt{3}$

(contd.)

Right-Triangle Formulas

4.	Side of square Diagonal of square	4. $s = \frac{1}{2}d\sqrt{2}$ $d = s\sqrt{2}$
5.	Altitude to hypotenuse Leg of right triangle	5. $\frac{p}{h} = \frac{h}{q}$, $h^2 = pq$, $h = \sqrt{pq}$ $\frac{c}{a} = \frac{a}{p}$, $a^2 = pc$, $a = \sqrt{pc}$ $\frac{c}{b} = \frac{b}{q}$, $b^2 = qc$, $b = \sqrt{qc}$

Coordinate-Geometry Formulas

1.	Midpoint M Distance P_1P_2 Slope of $\overleftrightarrow{P_1P_2}$	1. $x_M = \frac{x_1 + x_2}{2}$, $y_M = \frac{y_1 + y_2}{2}$ $d = \sqrt{(x_2 - x_1)^2 + (y_2 - y_1)^2}$ $m = \frac{y_2 - y_1}{x_2 - x_1}$, $m = \frac{\Delta y}{\Delta x}$, $m = \tan i$
2.	Slopes of parallels, L_1 and L_2 Slopes of perpendiculars, L_1 and L'	2. Same slope, m $mm' = -1$ $m' = -\frac{1}{m}$, $m = -\frac{1}{m'}$
3.	Equation of L_1, parallel to x-axis Equation of L_2, parallel to y-axis	3. $y = k'$ $x = k$
4.	Equation of L_1 with slope m and y-intercept b Equation of L_2 with slope m passing through the origin Equation of L_1 with x-intercept a and y-intercept b Equation of L_3 with slope m and passing through (x_1, y_1)	4. $y = mx + b$ $y = mx$ $\frac{x}{a} + \frac{y}{b} = 1$ $y - y_1 = m(x - x_1)$
5.	Equation of circle with center at origin and radius r	5. $x^2 + y^2 = r^2$

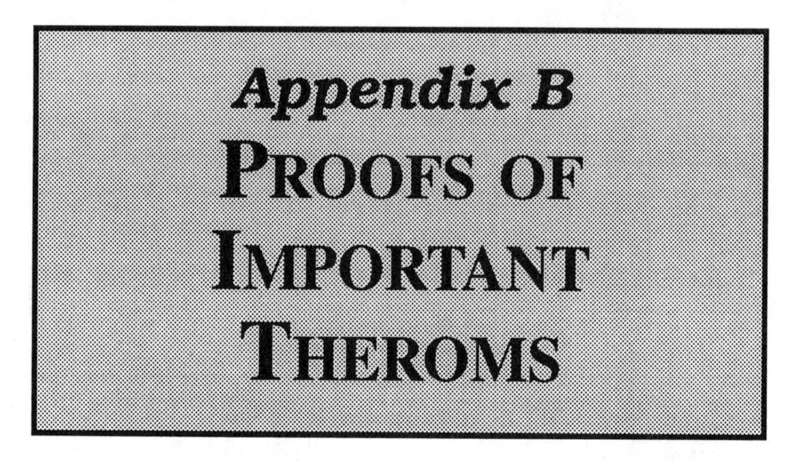

Appendix B Proofs of Important Theroms

The theorems and proofs given below are considered the most important in the logical sequence of geometry.

1. If two sides of a triangle are congruent, the angles opposite these sides are congruent. (Base angles of an isosceles triangle are congruent.)

Given: $\triangle ABC$, $\overline{AB} \cong \overline{BC}$
To Prove: $\angle A \cong \angle C$
Plan: When the bisector of the vertex angle is drawn, the angles to be proved congruent become corresponding angles of congruent triangles.

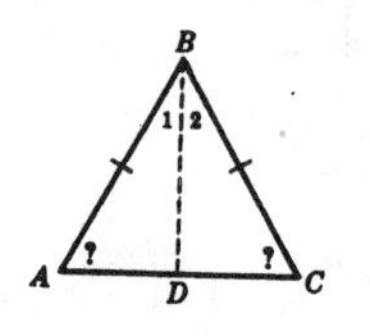

PROOF:

Statements	Reasons
1. Draw $\overline{BD}$ bisecting $\angle B$.	1. An angle may be bisected.
2. $\angle 1 \cong \angle 2$	2. To bisect is to divide into two congruent parts.
3. $\overline{AB} \cong \overline{BC}$	3. Given
4. $\overline{BD} \cong \overline{BD}$	4. Reflexive property
5. $\triangle ADB \cong \triangle BDC$	5. s.a.s. ≅ s.a.s.
6. $\angle A \cong \angle C$	6. Corresponding parts of congruent triangles are congruent.

2. The sum of the measures of the angles in a triangle equals 180°.

Given: $\triangle ABC$
To Prove: $m\angle A + m\angle B + m\angle C = 180°$
Plan: When a line is drawn through one vertex parallel to the opposite side, a straight angle is formed whose parts can be proved congruent to the angles of the triangle.

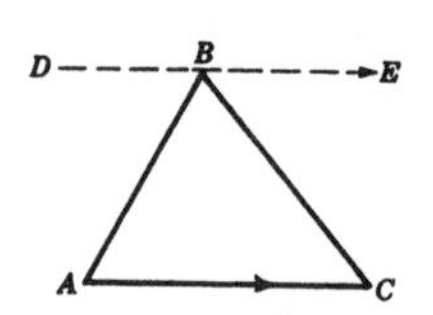

PROOF:

Statements	Reasons
1. Through B, draw $\overleftrightarrow{DE} \parallel \overleftrightarrow{AC}$.	1. Through an external point, a line can be drawn parallel to a given line.
2. $m\angle DBE = 180°$	2. A straight angle is an angle whose measure is 180°
3. $m\angle DBA + m\angle ABC + mCBE = 180°$	3. The whole equals the sum of its parts.
4. $\angle A \cong \angle DBA$, $\angle C \cong CBE$	4. Alternate interior angles of parallel lines are congruent.
5. $m\angle A + m\angle B + m\angle C = 180°$	5. Substitution Postulate

3. If two angles of a triangle are congruent, the sides opposite these angles are congruent.

Given: $\triangle ABC$, $\angle A \cong \angle C$
To Prove: $\overline{AB} \cong \overline{BC}$
Plan: When the bisector of $\angle B$ is drawn, the sides to be proved congruent become corresponding sides of congruent triangles.

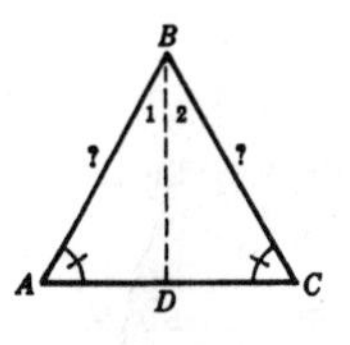

PROOF:

Statements	Reasons
1. Draw $\overline{BD}$ bisecting $\angle B$.	1. An angle may be bisected.
2. $\angle 1 \cong \angle 2$	2. To bisect is to divide into two congruent parts.
3. $\angle A \cong \angle C$	3. Given
4. $\overline{BD} \cong \overline{BD}$	4. Reflexive property
5. $\triangle BDA \cong \triangle BDC$	5. s.a.a. ≅ s.a.a.
6. $\overline{AB} \cong \overline{BC}$	6. Corresponding parts of congruent triangles are congruent.

4. Two right triangles are congruent if the hypotenuse and a leg of one are congruent to the corresponding parts of the other.

Given: Right $\triangle ABC$ with right angle at C
Right $\triangle DEF$ with right angle at F
$\overline{AB} \cong \overline{DE}$, $\overline{BC} \cong \overline{EF}$
To Prove: $\triangle ABC \cong \triangle DEF$
Plan: Move the two given triangles together so that $\overline{BC}$ coincides with $\overline{EF}$, forming an isosceles triangle. The given triangles are proved congruent by using Theorem 1 and s.a.a. ≅ s.a.a.

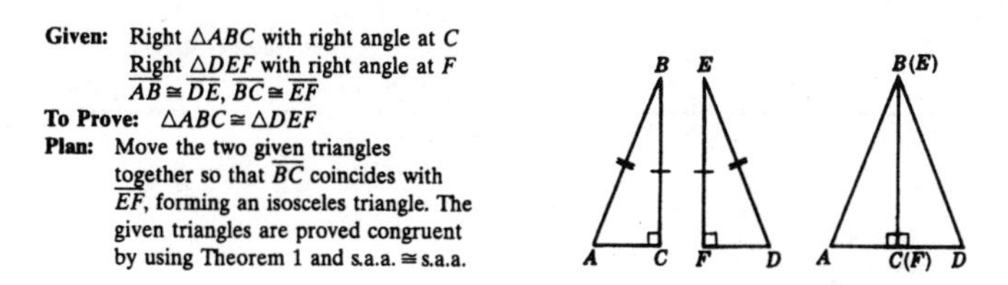

PROOF:

Statements	Reasons
1. $\overline{BC} \cong \overline{EF}$	1. Given
2. Move triangles ABC and DEF together so that $\overline{BC}$ coincides with $\overline{EF}$, and A and D are on opposite sides of $\overline{BC}$.	2. A geometric figure may be moved without changing its size or shape. Equal lines may be made to coincide.
3. $\angle C$ and $\angle F$ are right angles.	3. Given
4. $\angle ACD$ is a straight angle.	4. The whole equals the sum of its parts.
5. $\overline{AD}$ is a straight line segment.	5. The sides of a straight angle lie in a straight line.
6. $\overline{AB} \cong \overline{DE}$	6. Given
7. $\angle A \cong \angle D$	7. If two sides of a triangle are congruent, the angles opposite these sides are congruent.
8. $\triangle ABC \cong \triangle DEF$	8. s.a.a. ≅ s.a.a.

5. A diameter perpendicular to a chord bisects the chord and its arcs.

Given: Circle O, diameter $\overline{AB} \perp \overline{CD}$
To Prove: $\overline{CE} \cong \overline{ED}$, $\widehat{BC} \cong \widehat{BD}$, $\widehat{AC} \cong \widehat{AD}$
Plan: Congruent triangles are formed when radii are drawn to C and D, proving $\overline{CE} \cong \overline{ED}$. Equal central angles are used to prove $\widehat{BC} \cong \widehat{BD}$; then the Subtraction Postulate is used to prove $\widehat{AC} \cong \widehat{AD}$.

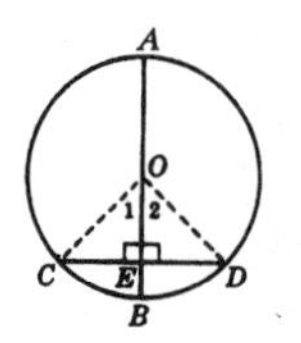

PROOF:

Statements	Reasons
1. Draw $\overline{OC}$ and $\overline{OD}$.	1. A straight line may be drawn between two points.
2. $\overline{OC} \cong \overline{OD}$	2. Radii of a circle are congruent.
3. $\overline{AB} \perp \overline{CD}$	3. Given
4. $\angle OEC$ and $\angle OED$ are right angles.	4. Perpendiculars form right angles.
5. $\overline{OE} \cong \overline{OE}$	5. Reflexive property
6. $\triangle OEC \cong \triangle OED$	6. hy. leg ≅ hy. leg
7. $\overline{CE} \cong \overline{ED}$, $\angle 1 \cong \angle 2$	7. Corresponding parts of congruent triangles are congruent.
8. $\widehat{CB} \cong \widehat{BD}$	8. In a circle, congruent central angles have congruent arcs.
9. $\widehat{ACB} \cong \widehat{ADB}$	9. A diameter bisects a circle.
10. $\widehat{AC} \cong \widehat{AD}$	10. In a circle, congruent arcs are equal arcs; Subtraction Postulate

6. An angle inscribed in a circle is measured by one-half its intercepted arc.

Case I: The center of the circle is on one side of the angle.

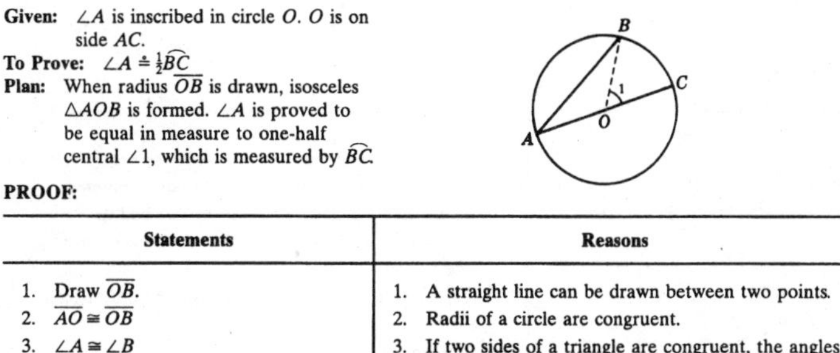

Given: $\angle A$ is inscribed in circle O. O is on side AC.
To Prove: $\angle A \doteq \frac{1}{2}\widehat{BC}$
Plan: When radius $\overline{OB}$ is drawn, isosceles $\triangle AOB$ is formed. $\angle A$ is proved to be equal in measure to one-half central $\angle 1$, which is measured by $\widehat{BC}$.

PROOF:

Statements	Reasons
1. Draw $\overline{OB}$.	1. A straight line can be drawn between two points.
2. $\overline{AO} \cong \overline{OB}$	2. Radii of a circle are congruent.
3. $\angle A \cong \angle B$	3. If two sides of a triangle are congruent, the angles opposite these sides are congruent.
4. $m\angle A + m\angle B = m\angle 1$	4. In a triangle the measure of an exterior angle equals the sum of the measures of the two adjacent interior angles.
5. $m\angle A + m\angle A = 2m\angle A = m\angle 1$	5. Substitution Postulate
6. $m\angle A = \frac{1}{2}m\angle 1$	6. Halves of equals are equal.
7. $\angle 1 \doteq \widehat{BC}$	7. A central angle is measured by its intercepted arc.
8. $\angle A \doteq \frac{1}{2}\widehat{BC}$	8. Substitution Postulate

Case II: The center is inside the angle.

Given: $\angle BAC$ is inscribed in circle O. O is inside $\angle BAC$.
To Prove: $\angle BAC \doteq \frac{1}{2}\widehat{BC}$
Plan: When a diameter is drawn, $\angle BAC$ is divided into two angles which can be measured by applying Case I.

PROOF:

Statements	Reasons
1. Draw diameter $\overline{AD}$.	1. A straight line may be drawn between two points.
2. $\angle BAD \doteq \frac{1}{2}\widehat{BD}$, $\angle DAC \doteq \frac{1}{2}\widehat{DC}$	2. An inscribed angle is measured by one-half its intercepted arc if the center of the circle is on one side.
3. $\angle BAC \doteq \frac{1}{2}\widehat{BD} + \frac{1}{2}\widehat{DC}$ or $\angle BAC \doteq \frac{1}{2}(\widehat{BD} + \widehat{DC})$	3. If equals are added to equals, the sums are equal.
4. $\angle BAC \doteq \frac{1}{2}\widehat{BC}$	4. Substitution Postulate

Case III: The center is outside the angle.

Given: $\angle BAC$ is inscribed in circle O. O is outside $\angle BAC$.
To Prove: $\angle BAC \doteq \frac{1}{2}\widehat{BC}$
Plan: When a diameter is drawn, $\angle BAC$ becomes the difference of two angles which can be measured by applying Case I.

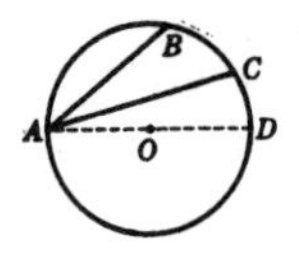

PROOF:

Statements	Reasons
1. Draw diameter $\overline{AD}$.	1. A straight line may be drawn between two points.
2. $\angle BAD \doteq \frac{1}{2}\widehat{BD}$, $\angle CAD \doteq \frac{1}{2}\widehat{CD}$	2. An inscribed angle is measured by one-half its intercepted arc if the center of the circle is on one side.
3. $\angle BAC \doteq \frac{1}{2}\widehat{BD} - \frac{1}{2}\widehat{CD}$ or $\angle BAC \doteq \frac{1}{2}(\widehat{BD} - \widehat{CD})$	3. If equals are subtracted from equals, the differences are equal.
4. $\angle BAC \doteq \frac{1}{2}\widehat{BC}$	4. Substitution Postulate

7. An angle formed by two chords intersecting inside a circle is measured by one-half the sum of the intercepted arcs.

Given: $\angle 1$ formed by chords $\overline{AB}$ and $\overline{CD}$ intersecting at point E inside circle O
To Prove: $\angle 1 \doteq \frac{1}{2}(\widehat{AC} + \widehat{BD})$
Plan: When chord $\overline{AD}$ is drawn, $\angle 1$ becomes an exterior angle of a triangle whose nonadjacent interior angles are inscribed angles measured by $\frac{1}{2}\widehat{AC}$ and $\frac{1}{2}\widehat{BD}$.

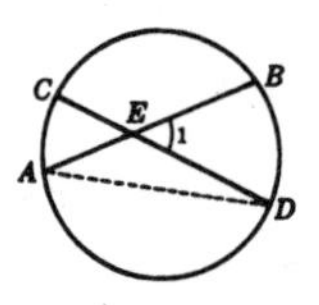

PROOF:

Statements	Reasons
1. Draw $\overline{AD}$.	1. A straight line may be drawn between two points.
2. $m\angle 1 = m\angle A + m\angle D$	2. The measure of an exterior angle of a triangle equals the sum of the measures of the nonadjacent interior angles.
3. $\angle A \doteq \frac{1}{2}\widehat{BD}$, $\angle D \doteq \frac{1}{2}\widehat{AC}$	3. An angle inscribed in a circle is measured by one-half its intercepted arc.
4. $\angle 1 \doteq \frac{1}{2}\widehat{BD} + \frac{1}{2}\widehat{AC} \doteq \frac{1}{2}(\widehat{BD} + \widehat{AC})$	4. Substitution Postulate

8(a). An angle formed by two secants intersecting outside a circle is measured by one-half the difference of its intercepted arcs.

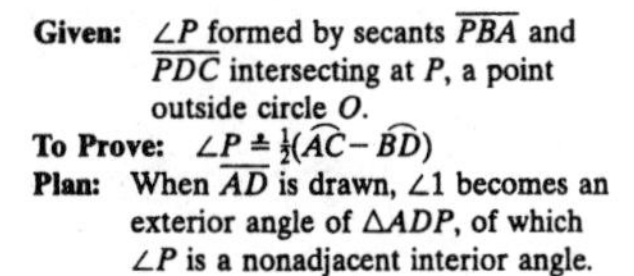

Given: $\angle P$ formed by secants $\overline{PBA}$ and $\overline{PDC}$ intersecting at P, a point outside circle O.
To Prove: $\angle P \doteq \frac{1}{2}(\widehat{AC} - \widehat{BD})$
Plan: When $\overline{AD}$ is drawn, $\angle 1$ becomes an exterior angle of $\triangle ADP$, of which $\angle P$ is a nonadjacent interior angle.

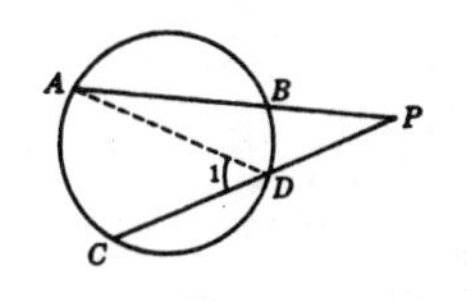

PROOF:

Statements	Reasons
1. Draw $\overline{AD}$.	1. A straight line may be drawn between two points.
2. $m\angle P + m\angle A = m\angle 1$	2. The measure of an exterior angle of a triangle equals the sum of the measures of the nonadjacent interior angles.
3. $m\angle P = m\angle 1 - m\angle A$	3. Subtraction Postulate
4. $\angle 1 \doteq \frac{1}{2}\widehat{AC}$, $\angle A \doteq \frac{1}{2}\widehat{BD}$	4. An angle inscribed in a circle is measured by one-half its intercepted arc.
5. $\angle P \doteq \frac{1}{2}\widehat{AC} - \frac{1}{2}\widehat{BD}$ or $\angle P \doteq \frac{1}{2}(\widehat{AC} - \widehat{BD})$	5. Substitution Postulate

8(b). An angle formed by a secant and a tangent intersecting outside a circle is measured by one-half the difference of its intercepted arcs.

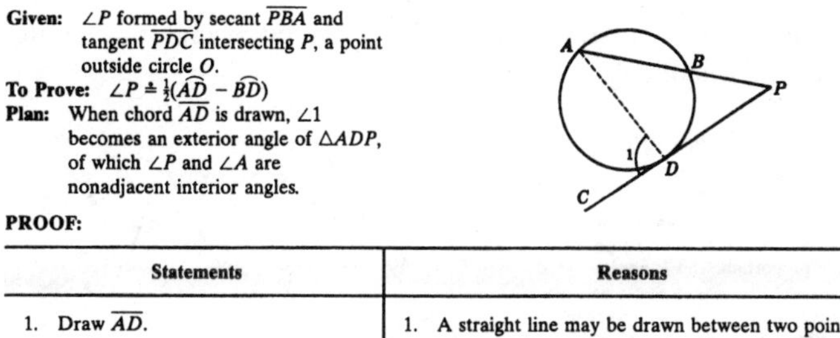

Given: $\angle P$ formed by secant $\overline{PBA}$ and tangent $\overline{PDC}$ intersecting P, a point outside circle O.
To Prove: $\angle P \doteq \frac{1}{2}(\widehat{AD} - \widehat{BD})$
Plan: When chord $\overline{AD}$ is drawn, $\angle 1$ becomes an exterior angle of $\triangle ADP$, of which $\angle P$ and $\angle A$ are nonadjacent interior angles.

PROOF:

Statements	Reasons
1. Draw $\overline{AD}$.	1. A straight line may be drawn between two points.
2. $m\angle P + m\angle A = m\angle 1$	2. The measure of an exterior angle of a triangle equals the sum of the measures of the nonadjacent interior angles.
3. $m\angle P = m\angle 1 - m\angle A$	3. Subtraction Postulate
4. $\angle 1 \doteq \frac{1}{2}\widehat{AD}$	4. An angle formed by a tangent and a chord is measured by one-half its intercepted arc.
5. $\angle A \doteq \frac{1}{2}\widehat{BD}$	5. An inscribed angle is measured by one-half its intercepted arc.
6. $\angle P \doteq \frac{1}{2}\widehat{AD} - \frac{1}{2}\widehat{BD}$ or $\angle P \doteq \frac{1}{2}(\widehat{AD} - \widehat{BD})$	6. Substitution Postulate

8(c). An angle formed by two tangents intersecting outside a circle is measured by one-half the difference of its intercepted arcs.

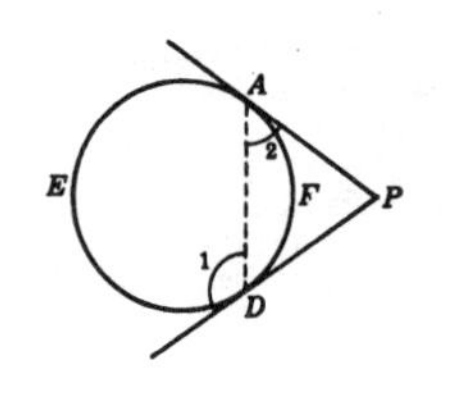

Given: $\angle P$ formed by tangents $\overline{PA}$ and $\overline{PD}$ intersecting at P, a point outside circle O.
To Prove: $\angle P \doteq \frac{1}{2}(\widehat{AED} - \widehat{AFD})$
Plan: When chord $\overline{AD}$ is drawn, $\angle 1$ becomes an exterior angle of $\triangle ADP$, of which $\angle P$ and $\angle 2$ are nonadjacent interior angles.

PROOF:

Statements	Reasons
1. Draw $\overline{AD}$.	1. A straight line may be drawn between two points.
2. $m\angle P + m\angle 2 = m\angle 1$	2. The measure of an exterior angle of a triangle equals the sum of the measures of the nonadjacent interior angles.
3. $m\angle P = m\angle 1 - m\angle 2$	3. Subtraction Postulate.
4. $\angle 1 \doteq \frac{1}{2}\widehat{AED}$, $\angle 2 \doteq \frac{1}{2}\widehat{AFD}$	4. An angle formed by a tangent and a chord is measured by one-half its intercepted arc.
5. $\angle P \doteq \frac{1}{2}\widehat{AED} - \frac{1}{2}\widehat{AFD}$ or $\angle P \doteq \frac{1}{2}(\widehat{AED} - \widehat{AFD})$	5. Substitution Postulate

9. If three angles of one triangle are congruent to three angles of another triangle, the triangles are similar.

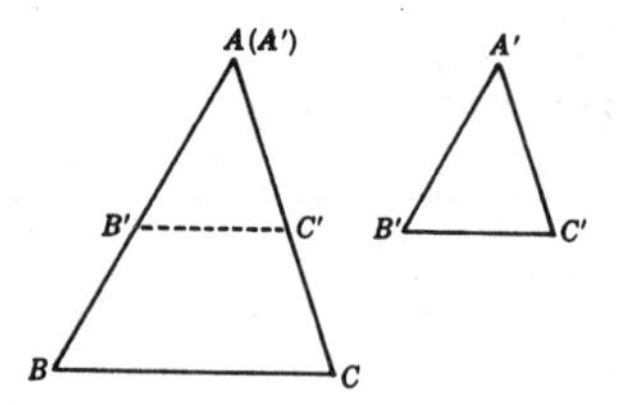

Given: $\triangle ABC$ and $\triangle A'B'C' \angle A \cong \angle A'$, $\angle B \cong \angle B'$, $\angle C \cong \angle C'$
To Prove: $\triangle ABC \sim \triangle A'B'C'$
Plan: To prove the triangles similar, it must be shown that corresponding sides are in proportion. This is done by placing the triangles so that a pair of congruent angles coincide, and then repeating this so that another pair of congruent angles coincide.

PROOF:

Statements	Reasons
1. $\angle A \cong \angle A'$	1. Given
2. Place $\triangle A'B'C'$ on $\triangle ABC$ so that $\angle A'$ coincides with $\angle A$.	2. A geometric figure may be moved without change of its size or shape. Equal angles may be made to coincide.
3. $\angle B \cong \angle B'$	3. Given
4. $\overline{B'C'} \parallel \overline{BC}$	4. Two lines are parallel if their corresponding angles are congruent.
5. $\frac{A'B'}{AB} = \frac{A'C'}{AC}$	5. A line parallel to one side of a triangle divides the other two sides proportionately

PROOF:

Statements	Reasons
6. In like manner, by placing $\triangle A'B'C'$ on $\triangle ABC$ so that $\angle B'$ coincides with $\angle B$, show that $\frac{A'B'}{AB} = \frac{B'C'}{BC}$	6. Reasons 1 to 5.
7. $\frac{A'B'}{AB} = \frac{A'C'}{AC} = \frac{B'C'}{BC}$	7. Things (ratios) equal to the same thing are equal to each other.
8. $\triangle A'B'C' \sim \triangle ABC$	8. Two polygons are similar if their corresponding angles are congruent and their corresponding sides are in proportion.

10. If the altitude is drawn to the hypotenuse of a right triangle, then (a) the two triangles thus formed are similar to the given triangle and to each other, and (b) each leg of the given triangle is the mean proportional between the hypotenuse and the projection of that leg upon the hypotenuse.

Given: $\triangle ABC$ with a right angle at C, altitude $\overline{CD}$ to hypotenuse $\overline{AB}$

To Prove: (*a*) $\triangle ADC \sim \angle BDC \sim \triangle ABC$

(*b*) $c:a = a:p$, $c:b = b:q$

Plan: The triangles are similar since they have a right angle and a pair of congruent acute angles. The proportions follow from the similar triangles.

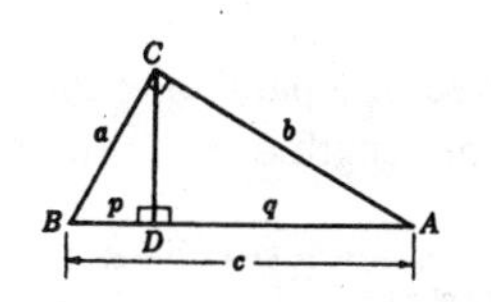

PROOF:

Statements	Reasons
1. $\angle C$ is a right angle.	1. Given
2. $\overline{CD}$ is the altitude to $\overline{AB}$.	2. Given
3. $\overline{CD} \perp \overline{AB}$	3. An altitude to a side of a triangle is perpendicular to that side.
4. $\angle CDB$ and $\angle CDA$ are right angles.	4. Perpendiculars form right angles.
5. $\angle A \cong \angle A$, $\angle B \cong \angle B$	5. Reflexive property
6. $\triangle ADC \sim \triangle ABC$, $\triangle BDC \sim \triangle ABC$	6. Right triangles are similar if an acute angle of one is congruent to an acute angle of the other.
7. $\triangle ADC \sim \triangle BDC$	7. Triangles similar to the same triangle are similar to each other.
8. $c:a = a:p$, $c:b = b:q$	8. Corresponding sides of similar triangles are in proportion.

11. The square of the length of the hypotenuse of a right triangle equals the sum of the squares of the lengths of the other two sides.

Given: Right $\triangle ABC$, with a right angle at C. Legs have lengths a and b, and hypotenuse has length c.
To Prove: $c^2 = a^2 + b^2$
Plan: Draw $\overline{CD} \perp \overline{AB}$ and apply Theorem 10.

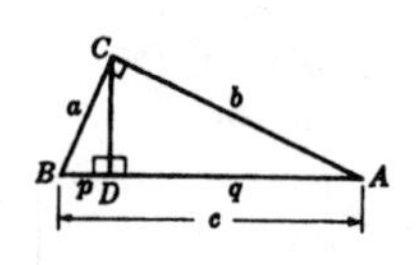

PROOF:

Statements	Reasons
1. Draw $\overline{CD} \perp \overline{AB}$.	1. Through an external point, a line may be drawn perpendicular to a given line.
2. $\frac{c}{a} = \frac{a}{p}, \frac{c}{b} = \frac{b}{q}$	2. If the altitude is drawn to the hypotenuse of a right triangle, either leg is the mean proportional between the hypotenuse and the projection of that leg upon the hypotenuse.
3. $a^2 = cp,\ b^2 = cq$	3. In a proportion, the product of the means equals the product of the extremes.
4. $a^2 + b^2 = cp + cq = c(p + q)$	4. If equals are added to equals, the sums are equal.
5. $c = p + q$	5. The whole equals the sum of its parts.
6. $a^2 + b^2 = c(c) = c^2$	6. Substitution Postulate

12. The area of a parallelogram equals the product of the length of one side and the length of the altitude to that side.

Given: $\square ABCD$, length of base $\overline{AD} = b$, length of altitude $\overline{BE} = h$
To Prove: Area of $ABCD = bh$
Plan: When a perpendicular is dropped to the base, extended, a rectangle is formed having the same base and altitude as the parallelogram. By adding congruent triangles to a common area, the rectangle and parallelogram are proved equal in area.

PROOF:

Statements	Reasons
1. Draw $\overline{CF} \perp \overleftrightarrow{AD}$ (extended).	1. Through an external point, a line may be drawn perpendicular to a given line.
2. $\overline{CF} \parallel \overline{BE}$	2. Segments perpendicular to the same line are parallel.
3. $\overline{BC} \parallel \overline{AD}$	3. Opposite sides of a parallelogram are parallel.
4. $\angle CFD$ and $\angle BEA$ are right angles.	4. Perpendiculars form right angles.
5. $BCFE$ is a rectangle.	5. A parallelogram having a right angle is a rectangle.

(Cont.)

PROOF:

Statements	Reasons
6. $\overline{AB} \cong \overline{CD}$, $\overline{CF} \cong \overline{BE}$	6. Opposite sides of a parallelogram are equal.
7. $\triangle ABE \cong \triangle CFD$	7. Hy. leg ≅ hy. leg
8. Area(quadrilateral $BCDE$) = area(quadrilateral $BCDE$)	8. Reflexive property
9. Area ($\triangle ABE$) + area (quadrilateral $BCDE$) = area($\triangle CFD$) + area(quadrilateral $BCDE$) or area(rectangle $BCFE$) = area($\square ABCD$)	9. If equals are added to equals, the sums are equal.
10. Area of rectangle $BCFE = bh$	10. The area of a rectangle equals the product of the lengths of its base and altitude.
11. Area of $\square ABCD = bh$	11. Substitution Postulate

13. The area of a triangle is equal to one-half the product of the length of one side and the length of the altitude to that side.

Given: $\triangle ABC$, length of base $\overline{AC} = b$, length of altitude $\overline{BD} = h$

To Prove: Area of $\triangle ABC = \frac{1}{2}bh$

Plan: Drawing $\overline{BE} \parallel \overline{AC}$ and $\overline{EC} \parallel \overline{AB}$ forms a parallelogram having the same base and altitude as the triangle. Then the area of the triangle is half the area of the parallelogram.

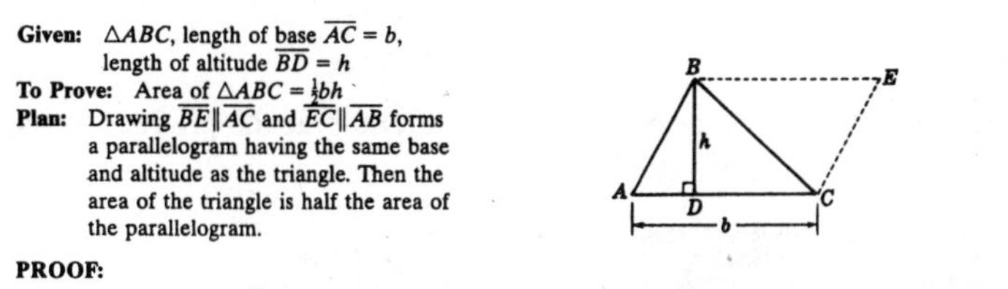

PROOF:

Statements	Reasons
1. Draw $\overline{BE} \parallel \overline{AC}$, $\overline{CE} \parallel \overline{AB}$.	1. Through an external point, a line may be drawn parallel to a given line.
2. $ABEC$ is a parallelogram with base b and altitude h.	2. A quadrilateral is a parallelogram if its opposite sides are parallel.
3. Area($\triangle ABC$) = $\frac{1}{2}$area ($\square ABEC$)	3. A diagonal divides a parallelogram into two congruent triangles.
4. Area($\square ABEC$) = bh	4. The area of a parallelogram equals the product of the lengths of its base and altitude.
5. Area of $\triangle ABC = \frac{1}{2}bh$	5. Substitution Postulate

14. The area of a trapezoid is equal to one-half the product of the length of the altitude and the sum of the lengths of the bases.

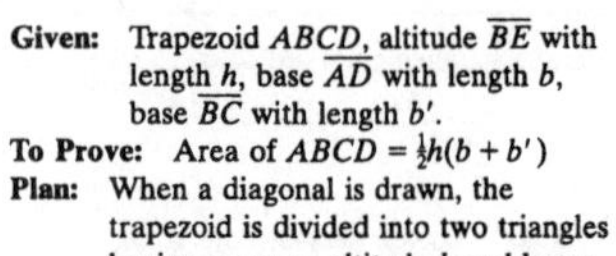

Given: Trapezoid $ABCD$, altitude $\overline{BE}$ with length h, base $\overline{AD}$ with length b, base $\overline{BC}$ with length b'.

To Prove: Area of $ABCD = \frac{1}{2}h(b + b')$

Plan: When a diagonal is drawn, the trapezoid is divided into two triangles having common altitude h and bases b and b'.

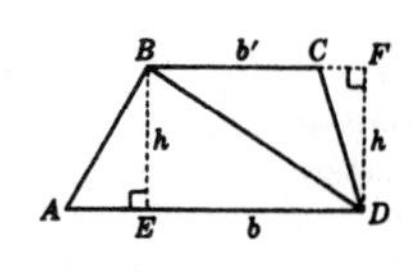

PROOF:

Statements	Reasons
1. Draw $\overline{BD}$.	1. A straight line may be drawn between two points.
2. Draw $\overline{DF} \perp \overrightarrow{BC}$ (extended).	2. Through an external point, a line may be drawn perpendicular to a given line.
3. $DF = BE = h$	3. Parallel lines are everywhere equidistant.
4. Area($\triangle ABD$) $= \frac{1}{2}bh$, Area($\triangle BCD$) $= \frac{1}{2}b'h$	4. The area of a triangle equals one-half the product of the lengths of its base and altitude.
5. Area of $ABCD = \frac{1}{2}bh + \frac{1}{2}b'h = \frac{1}{2}h(b + b')$	5. If equals are added to equals, the sums are equal.

15. The area of a regular polygon is equal to one-half the product of its perimeter and apothem.s

Given: Regular polygon $ABCDE \ldots$ having center O, apothem of length r, perimeter p

To Prove: Area of $ABCDE \ldots = \frac{1}{2}rp$

Plan: By joining each vertex to the center, congruent triangles are obtained, the sum of whose areas equals the area of the regular polygon.

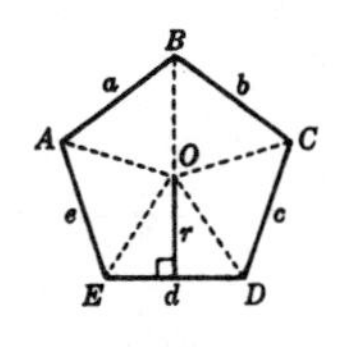

PROOF:

Statements	Reasons
1. Draw $\overline{OA}$, $\overline{OB}$, $\overline{OC}$, $\overline{OD}$, $\overline{OE}$,....	1. A straight line segment may be drawn between two points.
2. r is the altitude of each triangle formed.	2. Apothems of a regular polygon are congruent.
3. Area of $\triangle AOB = \frac{1}{2}ar$ $\triangle BOC = \frac{1}{2}br$ $\triangle COD = \frac{1}{2}cr$	3. The area of a triangle equals one-half the product of the length of its base and altitude.
4. Area of regular polygon $ABCDE \ldots$ $= \frac{1}{2}ar + \frac{1}{2}br + \frac{1}{2}cr + \cdots$ $= \frac{1}{2}r(a + b + c + \cdots)$	4. If equals are added to equals, the sums are equal.
5. $p = a + b + c + \cdots$	5. The whole equals the sum of its parts.
6. Area of $ABCDE \ldots = \frac{1}{2}rp$	6. Substitution Postulate

Index